BOILING LAVA UNIVERSE

Boiling lava Universe

Goran Mitic

Contents

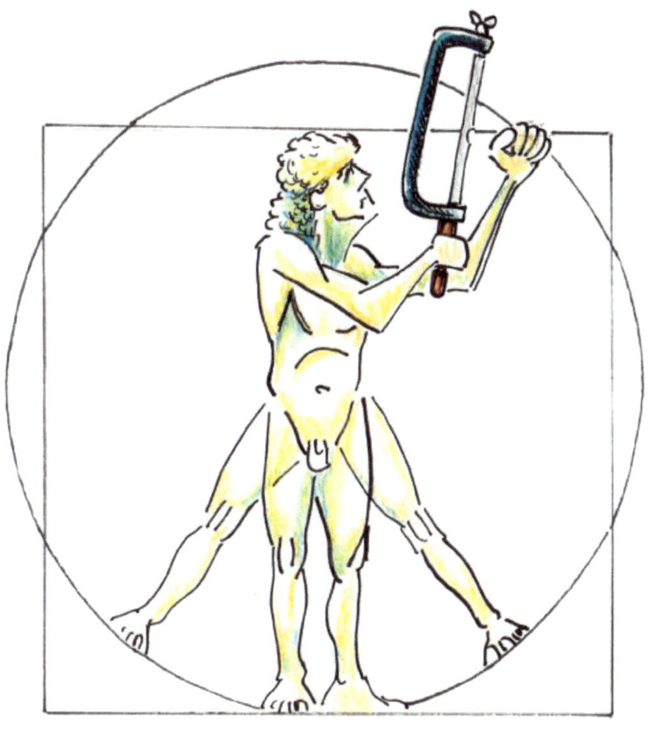

Lindqvist C.

HEART SCIENCE

Heart Science is the spiritual heart level of consciousness science. It is different in many ways from the mind level of science. The primary and most significant difference that heart science brings is the direction of our search for knowledge, for The Truth. The direction of the search in Heart science is inwardly oriented. Unlike the direction of the search in the mind science where it is outward. In heart science, we go inward to our spiritual heart, which is in the center of our chests, and we ask, with utmost sincerity, with the inner cry: What is it? What are we looking at? And then we keep silent. We keep waiting for the answer without thinking. When the answer comes from inside, we must use our minds to understand what the answer is. How it works? What is the composition? What is orientation and everything else to make in a system that stands for true knowledge and the truth? The consciousness of our spiritual heart is oneness and love for everything that exists. And everything that exists is One. This is why we cannot understand everything in our minds. Because our mind always needs two things to compare one with another. And because there are no two things, because there is only One, the mind does not have anything to compare with. Because of that, we must use our spiritual hearts to find answers. What mind science has conducted so far is a bunch of illog-

ical theories that are impossible to pass a logical examination. In cosmology, we miss 95% of the universe. We have dark matter. We have dark energy. We have Black Holes; we have Big Bangs.

These are all illogical and impossible-to-prove theories. Present mind science is the victim of our beliefs. Our beliefs are self-confirming. We see what we believe, and here we have a real problem. We look at our Sun; if we think it is a gas body made of hydrogen and helium, we see it that way. We understand the data we collect through our beliefs. When we look at the Sun by Heart Science, the spiritual level of our spiritual heart is visible, as if it is liquid boiling lava. So, it is not a matter of what we are looking at but the level of consciousness from which we are looking at a certain thing. There is no way that anybody can change anybody's belief system. We can only do that by going to our spiritual heart and asking. Is what I am talking about true? Is it the truth? The answer will come from within.

Due to the research achieved with the James Webb telescope and Hubble Telescope, we know our cosmology is wrong. Comparing the data telescopes collect and Mind Science present. Mind Science is inventing many new theories to escape the problems.

By pushing our minds further and further, mind science will create increasingly illogical and impossible theories. People are curious, so it is wrong to stop that, and we must switch to Heart Science. Otherwise, we will just continue going further and further in the wrong direction. For example, in 2023, 5.1 million science papers were published, and one-quarter of them couldn't be verified by other scientists so they are fake. Mind Science uses the method "fake it until you make it. "

How to become a Heart Scientist: We must learn to concentrate and meditate on our Spiritual Heart. This must be an everyday practice until we achieve mastery in that realm. After that, we will combine Heart Science and the level of the Mind, but we must remember and understand that Mind Science must be founded and guided by Heart Science. Otherwise, we will get back to the situation we are in now. The mind cannot go alone. It must be guided by the higher level of consciousness, our Spiritual Heart. Therefore, Heart Science must be a tutor, teacher, or leader to the Mind. The Mind's real purpose is to follow the Spiritual Heart and not to lead in any way.

Temperature Relativity

The concept of relativity is used in physics for all those values and notions that are, for any reason, variable, that is, do not support the same value. The popularity of the term 'relativity' has abruptly risen and stayed elevated since Einstein's theories of relativity appeared. We shall leave Einstein's theories of relativity for later because relativity is connected to speed. Here, I would like to talk about the relativity of values and notions regarding temperature. Let us see which things and how they change in nature with the temperature change.

The physical body of the basic notions in physics can be solid, liquid, or gas because those are the three essential states of aggregation we meet in our everyday lives. However, when we say a body, we first think of something with a stable shape and certain firmness. That is why we will start from that point, the solid body, that is, from the body in the solid state of aggregation.

When we see a solid body and follow what happens with its temperature changes, we first see that its dimensions change. With the increase in temperature, dimensions enlarge, and with the decrease in temperature, dimensions reduce. Briefly, „all bodies expand in heat and shrink in the cold; "thus, they change their volume (V). Then we will notice that the sole hardness of the body is changing. As the body's temperature rises, its hardness decreases, and we can more easily change its external shape. As the temperature decreases, its hardness increases and any attempt to reshape it is hardly possible, leading to cracking or breakage of the body.

Next, any further increase in the body's temperature leads to its melting, the transition from the solid into the liquid state of aggregation. Since the body has no specific shape, we need an open vessel made from solid material to preserve it in the liquid state. The hardness level is shallow, allowing us to reshape it by casting it into different molds and cooling it down, bringing it back into the solid state of aggregation. That is the essence of metallurgy. If we continue increasing the temperature of the liquid body, we will notice that it is starting to evaporate more and more until it starts boiling until the whole liquid evaporates rapidly, and we will have to keep our body in the gas state of aggregation in a new wholly sealed vessel that will have to be considerably more significant because this state implies much larger volume then the liquid. The gas state of aggregation implies complete occupation of the vessel's available volume and a certain amount of gas pressure on all vessels' walls. We can most easily overlook this change of aggregation states in our everyday lives.

By seeing the game of ice, water, and vapor. Liquids and gases can flow. Thus, we use the term 'fluids'; for the inner quality of hardness, we denominate 'viscosity.' With the increase in temperature, viscosity is decreased; with6 the decrease in temperature, viscosity is increased; warmer fluids flow more easily than colder ones. So, with a gas's temperature increase, the pressure of its action on the vessel's walls also increases. With the reduction of a gas' temperature, the pressure on the walls of the vessel is reduced in the first place. Condensation is what happens next, that is, restoring the body into the liquid state of aggregation and then hardening, that is, crystallization when the body is converted into the solid state of aggregation.

When people started studying the effects of electricity and magnetism, they also noticed that electrified bodies exposed to heat reduce their electricity until it is completely lost. When a magnetic body is heated, it reduces its magnetism until it is wholly lost.

With the beginning of using electric current for our needs, people met problems of its conduction through cables and issues of protecting themselves using isolators. We have set up that conductors have their resistibility and that they cause energy loss during its transport. It has also been found that the electric current flow through cables causes their heating, which causes the increase of resistibility, which is the increase of losses. That is why we must ensure there is no overheating of the cables because it can lead to the ignition of the installations and cause a fire and danger. However, cooling down of the cables reduces their resistivity. What is especially interesting is that at exceptionally low temperatures, the resistivity of the cables is completely gone, and there are no losses in the transmission of the electric current. Once set up, a current circuit is permanently kept, and that interesting phenomenon is called superconductivity.

As for the isolators, we have discovered that they protect us from the electric current very well when they are cool enough. If they, by any chance, get overheated, the isolators get breached; that is, above certain temperatures, they also become conductive. There is also a group of materials we call semiconductors, which are the basis of what we denominate as electronics. Their characteristics are very changeable with the change in the temperature, so we must take exceptional care about that.

What we can easily see is the flammability of varied materials. Upon heating to specific temperatures, some materials get inflamed and start burning, and that is why we call that temperature the ignition temperature. In principle, Gas is the most easily inflammable, followed by liquids and solid bodies. Humankind has been using fire for its own purposes for a long time, so we are all acquainted with lighting and burning a fire.

We all know that amongst the inflammable materials, there are those that incinerate quicker and some slower. Those materials that incinerate the quickest we call explosives, and we deal with them very carefully. Naturally, we deal carefully with all flammable materials.

In its essence, burning is a chemical process of oxidation and a gusty one. That means that the change in temperature changes the chemical properties of a material. Regarding the fact that there is a lot of free oxygen in the atmosphere, oxidation is an omnipresent process, but that process takes place differently in different temperatures. In chemistry, it is well known that temperature not only significantly changes the chemical properties of certain substances but also leads to their disintegration or their becoming more compound, which results in the generation of new substances.

All the changes caused by lowering or raising the temperature critically influence the living world and the possibility of its survival and development. Our own life is defined by the temperature of our bodies, and if there is severe raising or lowering, we die.

But not only living creatures lose their existence with the rise in temperature. If we continue heating any known material on and on, it will, after the gas state of aggregation, change into the state we call plasma. With matter in the state of plasma, there are no more atoms because they have reached their disintegration, that is, partial or complete ionization. With the high-temperature plasma, all electrons are detached from the nucleus. Chemical elements lose their existence upon the emergence of plasma.

I will also speak here about the emanation of a body in different temperatures. We all know that all bodies emit electromagnetic energy into their surroundings, no matter their temperature. The temperature of the body is the one that decides the wavelength of the dominant electromagnetic emanation that a body emits. The higher the temperature of a body is, the more dominant the emission of the short wavelength and the larger the energy is. Inversely, the lower the body temperature is, the more dominant the emission of the high wavelength and the lower the energy is.

People are generally able to differentiate one part of the spectrum with their eyes, and that part is called the visible part of the spectrum, or the visible light. Through our skin, we can also sense a part of the spectrum that we call heat or IR radiation, and a part of the spectrum called UV radiation.

By simply touching we can decide if one body is warmer than another one. If bodies are too warm, and we cannot touch them, we can simply just decide which body is warmer by bringing our hands nearer to a safe distance. The heating of metal objects, where making them red-hot can be seen with the naked eye, is particularly interesting. With the further heating of such bodies, they begin emitting red light, which is called red incandescence, and afterward, there is the emission of white light or white incandescence. Burning is an excellent example as well, that is, the flame. If we see the flame, let's say of a candle, we can see that there are certain areas that burn in different lights. Where the temperature is highest, the flame is the brightest, and where the temperature is lower, the flame is a bit darker, that is, more red. While burning different substances create flames of assorted colors, that is, temperatures, and they can be found in all the colors of a rainbow.

THE REVELATION OF ANTI-GRAVITY

While I lived on the slope of Chegar Hill, which is on the outskirts of the town of Nis, my wife and I regularly hiked to the top of that hill where there is a monument to the unique bravery of the Serbian rebels from the First Serbian uprising against the Turks. Not willing to surrender to the Turks, the leader of the Serbs, Stevan Sindjelic, shot a storage of gunpowder, thus blowing up both the Serbs and the Turks. Using the heads of the Serbian rebels the Turks built a Skull Tower at the outskirts of Nis to intimidate the Serbian people and the rebels. Serbs are a particular nation, and they do not allow themselves to be intimidated, so after the failure of the first Serbian uprising, they sparked off the Second uprising and managed to free themselves from the five centuries of slaving under the Turks.

In the close vicinity of the monument, there is a football field for the village's football club, and the whole hill is covered in vineyards and orchards. We regularly run around the football field.

On an August day in 1997, we were a bit late, so as we were running, dusk fell. A large pile of dried vine stick bunches was on a field behind the goal that stood closer to the monument. The villagers usually use those as fuel for distilling rakija (Serbian brandy) or as fuel for heating, especially for lighting fire, as they burn easily and rapturously. However, that pile was not prepared for transport but for burning at that spot. And, as we were about to finish running, and it was a bit dark, a villager lit the huge pile just as we were in the vicinity, and we watched the whole scene. Proceeding with our running, we turned our backs to the fire. When we ran around the opposite goal, we faced the fire again and saw a fascinating sight. The fire had already overtaken the whole pile and was reaching its maximum. The flame was 10 meters high, and it was illuminating the whole top of the hill. I had never seen a larger fire in my entire life. Totally fascinated by the sight, we stood in front of the fire and admired the magnificent sight.

I was delighted and happy as a child and there were no thoughts in my head at that moment – there was just a picture of the huge fire whose flames were rising so forcibly and quickly upward, at the same time narrowing towards the middle that was reaching the highest point. From the top of the flame into the darkness, at a great speed, sparks were flying out. That enormous force of heated air and particles was clearly revealing its tremendous speed of going up.

Suddenly there was a flash in my entire being and an inner voice.

"That is anti-gravitation!" Followed by a feeling of a current sliding up my spine from the bottom to the top of my head. My whole skin was in goosebumps, and all my hair bristled. That feeling was already well known to me because it was a companion to several revelations I had had before, except that this one had the largest intensity I had ever felt. My look at the fire was no longer the same. Now, I was watching anti-gravitation in action. I did not, not for a single moment, suspect of the veracity of the thought that had occurred in my head. Similar experiences to this one had already persuaded me of the veracity of ideas bought in such a manner. Instantaneously questions appeared – what and how is happening in combustion that was already beginning to decrease and soon ended before our eyes. After the appearance of the revelation of anti-gravitation, nothing in my life has been the same. That revelation completely seized me, and soon, my wife asked me what was going on with me. It was only then that I told her what had happened that night when we were watching the fire on Chegar Hill.

It is usual that people think that it is enough just for an idea to appear in your mind and that the problem is solved; anyway, I also used to think like that. However, the truth is reversed: when an idea appears in your mind, it means that a huge work has just begun and that there is a long and arduous work about the complete understanding of that idea, and then its verification everywhere and in every occasion, and in the end its implementation into the existing science.

Even though I was not fully aware of that, I at once embarked on the process of understanding the very idea. Previously, I thought that a person has an idea that he or she works on, but in time, I was assured of something completely different—the idea has a person who works on it. It is as if the ideas choose and take people through whose work they will materialize themselves and become known in a way they want to.

Mass Temperature Relativity

I started by analyzing the fire. The flame stretches from the bottom of the burning material's pile (because that is how a fire starts), and the higher the material piles, the higher the final height of the flame is. Above the top of the flame there is a part that is invisible, that is, transparent, and it is much shorter than the flame. Above the transparent part begins the zone of the visible smoke. In the beginning, the smoke is lighter, and with the increase in height, it grows darker and darker. With the increase in height, the smoke goes upward slower and slower, and at a certain point, it reaches its final height. Since it cannot go up after the final height, and because of the coming up of the now smoke from below, there is a radial spreading of the smoke cloud on that height and then it looks like a thick pancake. When the process of burning finishes, the formed smoke cloud hangs in the air at that maximal height and then slowly starts losing height, and finally, it falls to the ground, farther or closer to the place of burning, depending on the airflow.

Therefore, the visible effect of burning forms the tendency of the hot gas up to the final height and its decline to earth when it cools down. But let's analyze individual gas molecules that result from combustion (CO_2 + H_2O). Hot gas molecules emit electromagnetic radiation in IR (infrared) and the visible part of the spectrum, and we see those emissions as flame, lighter or darker. In that stadium, they are rapidly going upward. In the transparent zone the molecules have chilled a bit, enough not to emit visible light, but only IR radiation, and continue going upward rapidly. The beginning of smoke is formed of molecules that are already cooled down so that along with emitting IR radiation, they start absorbing the Sun's light and continue moving upward at a lesser speed. Upon reaching the final, that is, the maximal height, molecules are in the state where the amount of the emitted energy is the same as the amount of the absorbed energy, and they hang in the air for some time without vertical movement. As the process of cooling down of the molecules is continually in action, there comes a moment when they start falling slowly – obviously when their emitted energy is lessening. As the process of cooling down continues, the falling of the gas molecules speeds up and ends with their final fall to the ground, when the molecule's temperature assimilates with the outside temperature.

The logic of my thinking was like this. If the molecules of the hot gas that are characterized by elevated temperatures fly rapidly upward, and if that is anti-gravitation in action, that must mean that molecules in elevated temperatures have negative/repulsive mass. But, as they cool down while moving away from Earth, they start going upward slowly, which means that the negativity/repulsively of their mass is changing in a way that it is being reduced. The change in the distance from the mass centers of the Earth as a planet and the molecules of the gas cannot cause such changes in their mutual interaction because the entire process ends at a negligibly low height regarding the Earth's radius. When the molecules of gas reach the final height and start hanging in the air, that means that they lost the negative/repulsive character of their mass, that is, reached the massless state, and at that point, there is no interaction with the Earth, not anti- gravitational, nor gravitational. But their cooling down is happening continuously, and that is why they start having the mass of the attracting character, and they at once start falling towards the earth because of the establishment of the gravitational interaction with the Earth. The more they cool down, the faster they are falling toward the earth, which tells us that their attractive mass is changing quantitatively (by growing increasingly) with the lowering of the temperature. The molecules have the maximal attractive mass when their temperature is equalized with the temperature of the surrounding air, as they had the maximal negative/repulsive mass when their temperature was equalized with the temperature of the flame. The higher the combustion (fire) temperature is, the higher the maximal height of the gases will be before returning to Earth.

Does it make sense to talk about the temperature relativity of mass?

Well... yes! If temperature influences that many features of a matter, as I had already said, it makes sense to talk about it influences the feature we name mass. Therefore, it does make sense to talk about the temperature relativity of mass.

Temperature relativity of mass is such that with the warming of a body, the attractiveness of its mass decreases upon quantity until it is lost completely, let us say, reaches zero. That is a state when a feature we name mass is lost and the body is found in the massless state. That is also a state when the qualitative change of the body mass is performed. With further warming, the mass of the body becomes qualitatively negative/repulsive, and with the rise in the body temperature, quantitatively, the negativity / repulsively of the mass rises. That means that the feature we call mass changes with the change of temperature not only quantitatively but qualitatively as well.

Does it make physical sense to discuss negative/repulsive mass and anti-gravitation from the perspective of forces in nature?

Let us remind ourselves about what physics says about forces in nature. Until now, physics has defined four types of forces. Those are: strong, weak, electromagnetic, and gravitational forces. Strong nuclear forces are forced to work on the level of the atom nucleus between the protons and neutrons, and they manage the stability of matter. By their intensity, those are the strongest of all forces known to us, and by range, the smallest. Weak forces are those functioning on the level of an atom, and they manage the radio-active breakdown of matter. By their intensity, they are weaker than the nuclear or strong (thus the name), but they are still extraordinarily strong, and their range is larger than with the strong forces. Electromagnetic forces are easier for us to grasp because, in our everyday lives, we meet electricity and magnetism. The electromagnetic force is weaker than the weak force. However, its third place is not to be underestimated. The range of the electromagnetic forces is much larger than both the strong and the weak forces, and it is obvious. The closest experience to us is the gravitational force because it practically influences our lives and movements. That is the weakest force by the intensity of all forces but the most dominant force in the whole universe because its range is exceptionally large. Apart from being different quantitatively, these four forces are qualitatively different as well. How? Well, in such a way that strong, weak and electromagnetic forces appear as attractive and negative/repulsive, and the gravitational force appears only as attractive. Is the gravitational force an exception?

The temperature relativity of the mass is just the thing that introduces harmony amongst all forces by introducing the negative character of the gravitational forces, that is, anti-gravity. All forces now become attractive and repulsive, which we have been eagerly expecting and which seems so natural and logical to us.

So... The answer is yes! It does make physical sense to talk about negative/repulsive mass and anti-gravitation. It is just what is lacking in theory.

OBVIOUS EVIDENCE

When a person works on a new idea, apart from the tremendous enthusiasm that he/she is filled with, occasionally there are periods when doubt overcomes him/her and when he/she starts asking himself/herself if it is all a delusion or a serious mistake.

I asked myself those questions: am I not making a mistake? Am I not in delusion?

If mass temperature relativity is reality, then besides the fire, there must exist at least some other obvious evidence proving anti-gravitation in action. And thus, my observation of the world around us began. I suspected all vertical movements, upward or downward, as well as all processes where there is warming or cooling.

We live on the surface of the planet Earth in its air layer which we call atmosphere. We breathe the omnipresent air and feel its temperature or movement, although we cannot see it. So, let us 'see' what and how it is happening with this air that is in continuous movement. We are all aware of the fact that we learned as kids in primary school, which states:

„Warmer air is lighter, and it goes upward, and cool air is heavier, and it goes downward. "That is just what proves what I have said about mass temperature relativity. But let us go example by example.

When we see a closed-air system like our room, for example, then it is quite clear the coolest air is next to the floor and the warmest just below the ceiling. For those reasons, we put the heating objects that we use to heat our premises as low as possible so that they would heat the air equally by the volume. If we open the door or a window and rise from the floor upward with a lit candle or a lit lighter, we will assure ourselves that cool air enters the room from below and the warm air exits the room from above. That is how our room is getting cooler – from below and then upward we do feel the cool air on our feet first. The warm air that leaves the room continues its movement upward since there is no ceiling to prevent it from moving. If you do not believe me, heat up your oven and then open the door holding your hand above the stove. Do not put your face above the stove, for the stifling air might burn you.

If we want to cool ourselves in the summer, then we will put the cooling system close to the ceiling, because the cool air falling to the floor will best cool the air in the whole room by volume.

If we retry the experiment with the lit candle or a lighter with the fridge or the freezer door left ajar, we will notice that cool air is getting out of the object down and that the warm air is entering it above. Therefore, we have a completely different situation when we compare airing up warmed and cooled closed spaces. Why is that so?

When we heat the air in a closed space, the pressure increases in the upper part where the warm air is, and the pressure decreases in the lower part where the air is cold. Heated molecules of air whose mass has become negative/repulsive exert pressure on the upper surface of the closed space and pile up in the upper part, creating increased pressure, too. Because of the decrease in the number of molecules exerting pressure on them, the cool molecules slowly move apart, and below where the cool air occurs, the decrease of pressure.

When we cool the air in a closed space, the increase in the pressure arises in the lower part where the air is cooler and the decrease in the pressure in the upper part where the air is warmer. Cooled air molecules, whose mass has become even more attractive, exert pressure on the bottom surface of the closed space and pile up in the lower part, creating increased pressure. Because of the reduction of the molecules that are suppressing them, warmer molecules slowly move apart, and up where the air is warmer, and the decrease of the pressure arises.

Let us now see an open system such as the atmosphere of our planet. Earth's gravitation attracts all air molecules and thus keeps them around itself. We know that.

The air pressure at sea level is one atmosphere, and with the increase of height it decreases because the air is getting rarefied. But not even at that lower level on the Earth's surface is the pressure the same everywhere, and we have areas of increased or decreased air pressure, which brings about horizontal movement of air masses, that is, winds.

Why does it come to those differences in the air pressure? They arise because of the differences in the heating of various parts of the Earth's surface. Earth's surface is about one-third land and about two-thirds water. Land and water surfaces are getting warm in diverse ways. Even land surfaces get warm differently depending on their texture and appearance. Atmospheric air cannot get warm directly from the Sun's emission, but it is heated by the surface above which it is found. A warmer surface heats the air molecules more, and they go upwards, leaving the decreased air pressure near the ground. Cooler, or cold surface, cools the air molecules, and they fall, creating increased air pressure near the ground. Yachtsmen are real experts in catching the warm air currents that go upward, and they use them as elevators for lifting their yachts upward.

When people realized how the air moves, they started making flying objects called balloons. Around the balloon, they tied ropes that held the basket for the passengers or cargo were put, and below the balloon's gap there was a burner that heated the air within the balloon. The air in the balloon is heated by turning on the heater. The air then exerts pressure on the upper surface of the balloon thus lifting it upward. Turning off the heater and cooling down the air or letting out the warm air at the top of the balloon (which is conducted by opening the top), the pressure of the warm air on the upper balloon's face decreases, and the balloon loses its height and starts falling. That is how people, without even knowing what it is all about, started using anti-gravitation for flying.

Even better for observation than the air, is steam. We can see steam and follow its movement up or down, here, or there. Whether we are cooking in the kitchen or having a hot shower in the bathroom, we can notice the hot steam molecules' movement upward and fall of the cooled steam molecules. The same is happening in the atmosphere where steam can be seen in the shape of clouds. During the day, as the Sun heats, the clouds move across the sky carried by the winds, and when the Sun sets, they get cooler and fall toward the ground and then we say it is foggy. The whole story about climate and weather is based on the mass temperature relativity of the air and steam molecules. As we have seen with fire, that is, smoke, the same is with steam – there is a certain maximal height which the steam can reach and which, after all, depends on its starting temperature. Planes fly on heights that exceed the maximal height of clouds, that is, above the clouds, and that provides us with the opportunity to see the wonderful world of clouds from above. Observe it when you can fly. You will see places that look like springs that rise above the level of the clouds.

Television and movies present us, almost every day, with a number of explosions. They are of different origins, so we will analyze them one by one category.

The first category of explosions is provoked by an abrupt turning of chemical energy (atomic and molecule) into heat energy. All the materials that this can be provoked with are called classic explosives. The list of classic explosives today is exceptionally long, and people are continuously working on its extension.

Historically, people started out with gunpowder, that is, dynamite, then TNT, etc. The army industry inches research and creates increasingly powerful explosives that are then 'very efficiently' used in continuous wars. The idea that stronger explosives can bring the world closer to peace is completely wrong, extremely dangerous, and historically proven – a failure. Alas, what can we see if we closely see explosions of classic explosives? Now of explosion, a big fireball is generated; its dimensions depend on the type and the amount of the explosive in use. In the next moment, the ball starts rising, enlarging, and losing the fire glow by turning into a lighter or darker smoke cloud which is getting deformed because of its movement through the air. If we continue following the process until its end, we will see that the speed of going upward and enlarging is going to decrease and that a moment will come when that smoke cloud achieves its maximal size and, more importantly, its maximal height. After a shorter or longer hanging in the air, the cloud starts falling with the inevitable falling apart because of the action of the omnipresent air currents. The features of an explosion directly depend on the amount of released energy.

The second category of explosives, by their nature, is caused by an abrupt conversion of nuclear energy into heat energy by the process of fission and tearing apart the atomic nucleus. These materials are called fission-type nuclear explosives. There are only a few, but only one would be enough to face us with the possibility of self-extinction. People have come into possession of these explosives in the past century and developed destructive abilities up to unthinkable proportions. 'Nuclear mushroom' stands above mankind's head as a guillotine. In the very name – 'nuclear mushroom' lies the description of the process of the fission nuclear explosion. By quality it is identical to the description of the explosion of the classic explosive; the only difference is in quantity. The explosion ball is of much larger dimensions, as the explosion cloud and the maximal height of its centurion reaches about ten kilometers. Again, features depend on the kind and amount of the nuclear explosive, that is, the freed energy intensity.

The third category of explosives is, by its nature, caused by the abrupt conversion of nuclear energy into heat energy by the process of fusion or creating helium atoms by fusing hydrogen atoms. This kind of explosives I have isolated these on purpose because I will, in the further course of my presentation, assure you that it is not fusion here we are talking about but a whole new process that we have not even marked yet let alone understood. All the same, this category of explosives is the most powerful explosion that people can perceive. By their quality, they are like the earlier categories of explosives, and by their quantity, they exceed all earlier categories because the emitted energy is by far the largest.

With all explosions, it is in action clear that we clearly recognize the mass temperature relativity, solely that with explosions, unlike with fire, the entire process is finished in a single moment, which provokes the generation of the explosion ball. The explosion ball is generated because of the strong anti-gravitational influence of the over-heated molecules that appeared in the explosion, which are instantaneously, forcefully, and briskly getting away from each other. In the next moment that ball of over-heated molecules with negative/repulsive mass reverberates from the earth and goes upward until it cools down and stops reverberating with the earth when it reaches its maximal height. When it gets even cooler and the mass of its molecules becomes attractive, its falling will begin until all the products of the explosion fall to the ground, where their motion began.

Now, I will consider the process of combustion and explosion in the zero-gravity state. In the past century men went into outer space. When it became a routine and when people started feeling safe in the cosmos, the party started at once. Those who were spending a long time in orbit started celebrating their birthdays in the zero-gravity state, and since all that was transmitted by TV, we could see a lit birthday candle burning in the zero-gravity space. The flame in the zero-gravity state has the shape of a perfect ball. Why is it like that, when we all know that the shape of the flame on the Earth looks like a drop with its tip striving upward no matter how we hold the candle? We, on Earth, live under the continuous effects of gravity, and each flame, which is by its nature a gravitational occurrence, is striving opposite the center of gravitation. In the zero-gravity state, the flame, as an anti-gravitational occurrence without a center of gravity to reverberate from, reverberates only from itself and thus forms the shape of a perfect ball. Explosions taking place in outer space have the shape of the perfect ball, like the flame of a candle.

Explosions of nova's and supernovae have the shape of a ball, but we will analyze them in a later presentation.

A nice example for proving the previously said could be an incense stick in the zero-gravity state. On earth, the incense stick's smoke goes straight up in a straight line because that is an anti-gravitational occurrence. I have not had the chance to see a lit incense stick in the zero-gravity state, but I claim that its flame would expand like a perfect ball, enlarging its diameter. Let those who can perform this harmless experiment.

An exceptional example both for its relevance and size, as well as its duration and beauty, is Aboriginal bonfires. Australian natives, Aboriginals, believing that they had been paid a visit from outer space, have the custom of making a huge fire every year in a certain month and keeping it burning throughout the whole month, in order to say 'hello' to their visitors and show them that they have not forgotten about them. This belief seems naive and cute to us. Moreover, I used to think like this, but I assure you that what the Aboriginals are doing is neither naive nor cute but completely makes sense and is very efficient.

Cosmonauts who were flying in the Earth's orbit at the time of this Aboriginal ritual claimed that it was that fire that helped them to orient themselves about the location at night. By observing Australia from above, they could clearly see the Aboriginal fire from that height, and they said it was something really fascinating. They said that they had the clear impression that the blazes were reaching all the way to the orbit. The conclusion is that the Aboriginals know exactly how big the fire needs to be and how long it has to be burning in order to be able to leave the Earth's gravitational field and the peaks of the atmosphere in order to emit their light undisturbedly in the wanted direction. We must not forget that the

Aboriginals do that in a certain month of the year, which means that they always send their message to the same part of the starry sky.

Aboriginal fires are proof that only with fire can the Earth's gravity be surmounted because, with fires, there is no limit we call the maximal height. Hot molecules of the Aboriginal fire leave the Earth's gravitational field and those are really the first launchings of material from the Earth into the cosmos. Besides, all the launches performed today are with the help of combustion and fire.

As we are speaking of launching, it is interesting to re-mind that men have used a certain launching device for a long time. That device is called a chimney. As it is used as a thermal isolator, a chimney enables us to launch products of combustion as smoke, ashes, and soot to the maximal possible height, so that, carried by the wind, they would fall as far as possible from ourselves, even if it is only in our first neighbor's yard.

Let us see what happens with liquids. It is well known to us how we should heat liquids, from below, of course. Heated parts of the liquid come out upward to the surface, where they get cooled down and then sink to the bottom, where they get heated again, which again leads them to the surface. That is a perfect consistency in behavior, as well as with gases. Mass temperature relativity functions identically with all fluids. On occasions when we heat food up to the point of boiling (dishes, broths, soups, teas, coffees, and others.), when we can easily see the movement of the sole liquid as well as the movement of the steam, we observe the anti-gravitation in action like we observe gravitation all the time.

In the zero-gravity state, liquids form the shape of a ball, a larger or a smaller one, depending on their amount. If we could put a heater in the center of the liquid ball and heat the liquid, we would provoke circulation of the hot liquid from the center towards the surface in all directions. And if boiling of the liquid occurred, it would be visible all over the surface of the ball.

Finally, let us see what happens with solid bodies. To understand the principle of heat stretching through solid bodies more easily, we shall see the heating of solid bodies that are heat conductors, like metals.

If we take a bit thicker metal bar, let us say 30cm long and 2 up to 3cm in diameter, and hold it with our hands by its ends, and we put the middle on a small, preheated burner of our kitchen stove, its heating will begin. As a good heat conductor, the metal will be heated in all directions from the heat source, but by far the most vertical above the place of heat. That can be set up by touching it if we have not overheated the material or by bending the bar that will bend exactly at the vertical of the heating. The whole blacksmith's technology is based on this fact. So, the model of heat transmission by the vertical from below upward is kept with the solid bodies, too, no matter that in solid bodies, there is no internal movement of the matter as with the fluids, that is, the liquids and the gases.

And as we can clearly say at the end of this story, nature around us does not hide anything. It functions by its laws and we, with the development of our awareness and power, discover or find its laws one by one. The turn has come to gravity. But it takes a long series of questions with it and a lot of problems. I have started walking that road, step by step and I have come to the new physics. Now, I am taking you to see what it all looks like. Although to be honest, that process is an infinite story, and it will only last until I am authoring this book, and it is only then that a big new beginning will occur in understanding the world around us, as of ourselves, too.

Sun, Our Star

There are innumerable examples of red-hot bodies in a zero-gravity state; if you do not believe me, look at the sky when the Sun sets. Every star we can and cannot see in the sky is a red-hot body. All those stars we can ob- serve with our naked eyes because they are extremely far and the intensity of their radiation reaching us is very weak. But during the day we are shone upon by a star whose radiation intensity is so strong that we cannot look at it with our naked eyes because we would become blind. We can only watch its rising and setting. That star is remarkably close to us, and it provides us with pleasant heat and light; however, it is far enough in order not to turn us into ashes and dust. We consider that star – our star and we call it the Sun. People have, from their very genesis, been fascinated by the Sun. They followed its movement across the sky and oriented themselves by it in space and time. Then they noticed the annual cycle of the Sun's movement and started counting years; that is how the calendar was made. We have understood when spring comes and when we seed our plant cultures; when the winter comes – to prepare enough food and fuel for that period. Before that, we moved south in the autumn, as the birds do, and went back north in spring. We have understood that the temperature on Earth is directly dependent on the Sun's position and movement across the sky. During the day, when we expose our bodies to the Sun's rays, we clearly sense the Sun's warmth on our skin. A question immediately arises: 'What is the source of that vast energy that the Sun is radiating'?

When several centuries ago, people invented the telescope, and they started observing all the celestial bodies systematically. At night, they observed the stars and the planets, and during the day – the Sun. The development of astronomy has changed our conception of the cosmos. We have realized that the Earth revolves around its axis that the Moon revolves around the Earth, and that the Earth, together with the Moon, revolves around the Sun, which also revolves around its axis and around the center of our galaxy.

The discovery that the Sun's light consists of a number of lights in assorted colors (the colors of the rainbow) led to the development of spectral analysis and the creation of different devices for that purpose. We have learned how to use spectral analysis to determine not only the temperature of a body emitting light but its chemical structure, both qualitatively and quantitatively.

Spectral analysis of the Sun's light led to the conclusion that 71% of the Sun's mass is hydrogen (H_2) and 27% helium (He). Other chemical elements: O, C, Fe, N, and Ne, constitute a bit more than 1% of the Sun's mass. When the total number of atoms that constitute the Sun is observed, then 91.2% are hydrogen atoms and 8.7% helium atoms. Temperature of the Sun's surface is about 5800K.

Scientific analysis and calculations performed at the end of the 19th and beginning of the 20th centuries to discover the origin of the Sun's energy took this course.

The possibility that the Sun's energy originates from the exothermic chemical reactions with the present Sun's luminous- it leads to results that it is enough for the Sun to be shining for only about 30,000 years. That is, of course, an utterly unsatisfactory result, and therefore, that possibility was rejected. The possibility that the Sun's energy stems from gravitational compression led to the result of 16.5 million years. That was not a satisfactory result either, so that possibility was rejected, too. Today, it is considered that the energy emission by gravitational compression is dominant only at the early and late stages of the evolution of all stars, thus, the Sun, too.

The possibility that the Sun's source of energy is radioactive decomposition, was also rejected because of its inadequacy.

The idea of hydrogen fusion was presented by Sir Edington in 1920. Theoretically calculating, he established that upon the 4 H nuclei cohesion, energy of 7MeV per nucleon is extracted to the nucleus of them. In 1938 Weizsäcker identified the possibility of fusion reactions of H2 into He through a proton and carbon-nitrogen cycle. In 1939, Bete and Critchfield established, by detailed calculations, that hydrogen fusion 'burning' guarantees enough energy for the Sun's luminosity for 10 billion years. That result finally satisfied the scientists. That is how the idea of the fusion of H into He became generally accepted as the source of the Sun's energy.

That led to the conclusion that the Sun is a gas sphere in mechanical balance; that is, its own force of gravity, tending to compress the star, is balanced by the force of the gas pressure, which tends to spread it away.

Of course, for the fusion to be happening in the Sun's center, an extremely high temperature is necessary. EM radiation we see from the Sun stems from a relatively thin surface layer. The huge thickness of the Sun's matter and its state led to its being almost opaque, even for the hardest gamma and X-ray radiation coming from the inside of the Sun. For those reasons, the inside of the Sun is not accessible to the observer, and it is judged based on theoretical models.

Standard Model Of The Sun

This standard model that is, with certain modifications and corrections, scientifically acceptable today, was given by Mr. Sears. It was built for stars with mass, radius, glow, and constitution that match the Sun. According to this model, the inside of the Sun is made from a core (a zone of fusion reactions), a radiation zone, and a convective zone. In the radiative zone, energy made in the core is transferred toward the outer layers by radiation. In the convective zone, the principal mechanism of energy transmission is convection, that is, matter flow.

The standard model (SM) supposes that in the center of the Sun, the temperature is 15 million degrees, and the density is 150000 kg/m3. Even though the word is about huge density and pressure, it is nonetheless considered that because of the elevated temperature, the substance is in the state of the completely ionized gas plasma that can be treated as a perfect gas.

SM is in concordance with theories about energy production in the Sun, and in that sense, it corresponds very well with the direct observations. To explain the results of precise measurements in all parts of the EM spectrum and the corpuscle radiation, the standard model has been changed several times, but its basic preferences are still valid in scientific circles. That is how, for example, today it is considered- that the core temperature is a bit lower than the one predicted by the model and that it is about 14 million degrees.

Sun's core, according to SM: In the center of the Sun, there is a compact core, which comprises about 60% of the Sun's mass. Its dimensions are r=0,25 R☉, which means that it takes only about 1.6% of the Sun's volume. To get fusion reactions, it is necessary for the atomic nuclei to get within a reach of less than 10^{-15}m. Then, a strong, attractive nuclear force starts acting between them. However, to bring the particles to such small distances, the strong coulomb force of bouncing off of the electrifications having the same name (which is, to that extent, the smaller the distances between the particles are) has to be overcome. One of the possibilities is that the particles are moving at great thermal speeds of more than several hundred kilometers per second. Such thermal speeds can be realized in temperatures that are of the order size 10^7K. If the thermal speeds are low, the particles will diffuse before they get to the distance where the attractive nuclear force becomes stronger than the bouncing coulomb force. The high internal energy of the Sun is initially provided by the powerful gravitational force that is a consequence of the great Sun's mass. It compresses gas and that is why it gets heated.

SM has determined the temperature for the Sun's core that enables running of the fusion reactions. Temperature of 15 million degrees, to which the Sun's core is heated according to SM, is insufficient for all the existing particles to inter-react in fusion. Namely, in clashes particles are usually diffused, and only a few of them enter fusion reactions. Plasma in the core is treated as an almost ideal gas so that at the end of Maxwell's classification of particles according to speed, there are very few protons that can realize fusion reactions. However, thanks to the quant effect of tunneling, enough particles overcome the electro-reflecting barrier, entering a nuclear reaction, even at lower temperatures.

Basic fusion reactions in the Sun's core are conducted in two cycles: proton-proton (P-P), which is dominant, and carbon-nitrogen cycle (C-N). Both cycles release the same energy, about

26.72 Mev per He-formed nucleus. Neutrinos occur in fusion reactions. They isolate about 2% of the freed energy in the P-P cycle and about 7% of energy in the C-N cycle. Today, several especially important and expensive experimental systems for the detection of solar neutrino function are available. The results of the measurements are unexpected for the astrophysicists: The number of detected neutrinos is less than anticipated based on the SM.

According to these calculations, when the Sun is more than 9 billion years old, all the supplies of hydrogen will be consumed and transformed into helium, and the zone of hydrogen fusion will start transferring towards the outer areas to the layer encircling the nucleus. This area will expand until it gets to the area in which the temperatures are lower than ten million degrees. Then it will come to the extinction of the hydrogen fusion. At the same time, the Sun's nucleus, rich in helium, will get compressed by the action of its own gravity. That will lead to the growth of pressure and temperature and the creation of new conditions for the starting of fusion reactions of helium nuclei. Carbon and oxygen nuclei will be formed in these reactions, and that will be followed by freeing energy.

Under the influence of the fusion reactions of helium in the nucleus and hydrogen in the thin layer far from the nucleus, the solar mantle will 'inflate', which will lead to the gradual enlargement of the solar radius. During this, which will last approximately five hundred million years, the Sun will turn into a red giant. It will then 'swallow' its system of planets, and the effective temperature of its 'surface' will become lower. Then, a short phase of fast fusion combustion of the remaining helium and the heavier elements (of about 50 million years) will ensue. During this phase of the Sun's evolution in its core only carbon and oxygen will be found. The inside of the Sun will continue its further collapse, which was only temporarily stopped by helium fusion. The core temperature will rise again, but it will not enable further fusion reactions. The atmosphere of the Sun will expand a bit more. Sun will start pulsating lightly to expand and compress, with periods of about several thousands of years.

Finally, this phase of evolution will end by casting the Sun's atmosphere in the shape of one or two expanding films. In their center, a core that will intensely emit ultraviolet radiation will remain. In that way, the Sun will turn into a planetary haze in whose center a white slowly-cooling dwarf will be found. After several billions of years of cooling down, the Sun will turn into a dark, brown dwarf—the ending stadium of its evolution.

Sun's radiative zone occupies the area of 0.25 – 0.85 Ro from the center of the Sun. In the radiation zone, as well as in the core, energy is transferred towards the outer layers by radiation. Since in the radiation zone there are no fusion reactions, there is no 'cumulation' of He, so the mass percentage of H_2 is double compared to the core. At the beginning of the radiation zone, T is about $7x10^6$K K, and at the end, it is about $2x10^6$K.

The sun's convection zone spreads to the area from the upper boundary of the radiation zone to the photosphere, that is, the surface of the Sun, which means that its thickness is between 150,000 and 200,000 km. In this zone, the dominant transfer of energy is performed by convection, which is stance flow. This zone is of great significance before all, because processes in it determine the characteristics and the behavior in the external parts of the star (the origination and the variations of the local magnetic fields, activity, heating of the upper layers of the atmosphere, and so on). In convective layers there is movement of large mass of the Sun's substance, where the warmer mass goes up towards the surface, as the cooler mass lowers towards the deeper layers. Gas that has erupted on the Sun's surface loses its energy by radiating, cools down, and then sinks again deeper and deeper into the warmer layers of the convective zone. By going down, the gas gets heated again, and the process of circulation goes on.

The speed of the convective movement at the surface layer of the Sun reaches 2-3 km/s. At the beginning of the convective layer, the temperature is 2×10^6 K, and at the surface, in the photosphere, it is about 5800K.

At the horizontal section, convective cells are almost hexagonal. In their center, the substance goes up, and at the periphery, it lowers towards the deeper layers.

Substance movement in the highest layers of the convective zone leads to the occurrence of granulation in the photosphere, acoustic perturbations, and gas oscillation in the Sun's atmosphere and, through them, to the heating of its higher layers. That is, in brief, the essence of the SM Sun. As we can see, it is already troubled by different problems and unanswered questions. Let's see what happens when we incorporate the mass temperature relativity and the anti-gravitation into the analysis.

THERMO NUCLEAR FUSION IS IMPOSSIBLE!

All that physics has done so far, including astrophysics too, is based on the theory that there is only and solely gravitation for mass interactions. When we enrich the mass interactions with anti-gravity, which is a natural course of action, everything will change significantly. How?

The first and basic conclusion that we reach is that thermonuclear fusion, or hot fusion, is impossible!

It is impossible for the hydrogen nuclei to fuse into helium because next to the Coulomb bouncing, they bounce off anti-gravitationally, too. At the temperature of 15×10^6K, the negative mass of the H nuclei is so large that there is no possibility of their fusion. With the rise of the temperature, that is, thermal speeds, the situation gets even worse for the possibility of fusion.

Fusion is possible at exceptionally low temperatures when the attractiveness of the atomic masses goes so high that it overcomes the force of their Coulomb bouncing off. Therefore, nature allows only cold fusion.

This claim is far too radical, and it demands at least some experimental proof. Is there any such evidence?

Of course, there is. Those are multi-decade tries to achieve controlled thermonuclear fusion in earthly conditions.

What a clever idea! Achieve thermonuclear fusion in earthly conditions and solve the problem of energy on the planet forever. An attempt whose aim justifies all material means and intellectual effort invested. The source of unlimited and pure energy is not only worthy of glory but is also financially completely tempting. Americans and Russians have (a few decades ago) started the realization, both in their own ways.

Americans named their project 'Shiva', after the god Shiva from the Hindu Holy Trinity. Their concept was to use immensely powerful lasers from different sides to hit a small ball filled with hydrogen. Regardless of all their effort, the enhancement of the laser power, and finally, expenditure of all financial means, there was no expected result.

Russians named their project 'tokamak', which is the shortened name of their experiment. Their concept was to, by using powerful magnetic fields, keep high-temperature plasma in the shape of one ring long enough until conditions for fusion appear. With the increase in the temperature, that plasma ring would always fall apart before there were any expected results. All the devoted effort, as well as the enhancement of the power of the magnetic fields, along with all the financial means invested, did not lead to the expected results.

There were no results, and there will be no, because both parties, led by an illusion that arose because of the deficiencies in the theory of natural forces, have been trying to conduct something that is not possible.

All further attempts to achieve thermonuclear or 'hot' fusion are doomed to fail and represent a futile waste of massive amounts of money and scientific potential.

But, what about the H bomb? Have we not conducted the uncontrolled thermonuclear fusion in earthly conditions in the H-bomb? The answer is: no, we have not!

We have not conducted thermonuclear fusion, uncontrolled, in the so-called H-bomb, and what is happening upon the explosion of the H-bomb, we will yet have to discover.

Our Solar System

This chapter is about our own Solar System. Looking further away from the Sun, we see four planets: Mars, Venus, Earth, and Mars. We call them rocky planets. We know that they are made of rock and have a rocky surface. And then they have lava below the surface. When we go further from the orbit of Mars. Between the orbit of Mars and the orbit of Jupiter, we see the asteroid belt. It's donat-shaped, full of asteroids and rocks of all kinds. When we look at the orbit of Jupiter, we see that there are two clouds even before Jupiter and after Jupiter. Also, rocks, one of the clouds we named Greeks. The other cloud we named Troyon's. But then we have the most prominent significance in our solar system: Jupiter, Saturn, Uranus, and Neptune, like gas planets.

We will see later that it is impossible because they will not be able to keep their orbiting satellites in orbit, not to keep themselves in a vacuum of space. And then, after the orbit of Pluto, which is who has lost the position of the planet, we see the Kuiper belt, its vast, enormous again donate shape. And it's enormously big. But if we go further, we see that the Kuiper belt is a tiny spot in the Oort cloud, and the Oort cloud is also full of rocks—comets, asteroids, meteors, rocks, and rocks. So, the pattern of the Solar System is: rocky planets, the asteroid belt between the orbit of Jupiter and

Mars, then four big planets with rocky satellites, which indicates that they are rocky as well, the Kuiper belt, the big one after the orbit of Pluto and everything is in Oort cloud, a vast giant sphere of rocks.

So, where is the logical problem? We have the logical problem because, in all the rocks around the Sun, we consider that the mother star, our Sun, is a gas body made of hydrogen and helium. So, how can a star, which cannot make any heavier element than helium, produce all these rocks around her? So, that is a logical problem we must solve. It is not possible to be like that.

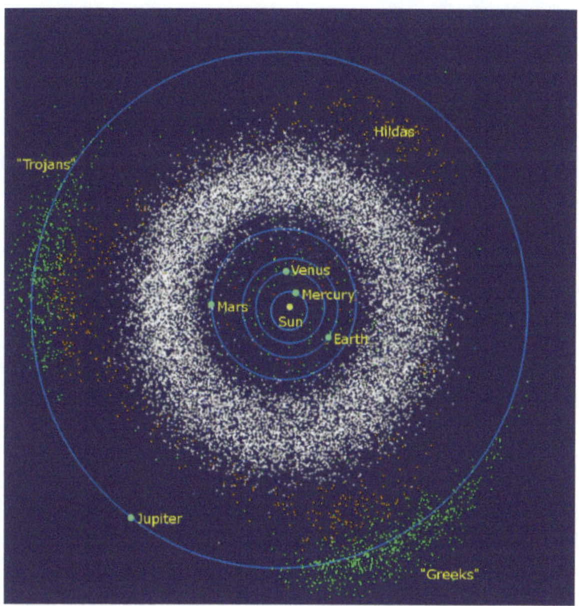

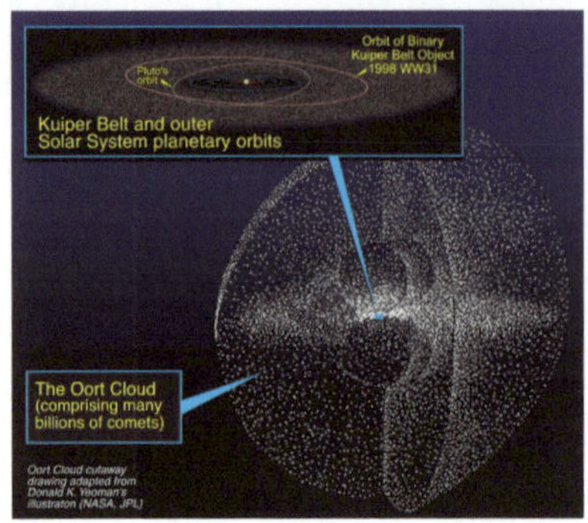

Orbit of Binary
Kuiper Belt Object
1998 WW31
Pluto's
orbit
Kuiper Belt and outer
Solar System planetary orbits
The Oort Cloud
(comprising many
billions of comets)
Oort Cloud cutaway
drawing adapted from
Donald K. Yeoman's
illustraton (NASA, JPL)

Boiling Lava Star Model

If TN fusion is impossible, then we must reopen the source of the Sun's energy, which is the fundamental astrophysics question about the origin of all stars as for the SM of the Sun, it was a complete disaster, and that is why it is necessary to create and think of a new model of the Sun that, because of the introduction of the anti-gravity, could be named the anti-gravitational model of the Sun.

In my antigravitational model of the Sun (AMS), I will begin analyzing what we see on the Sun's surface. The Sun's surface is quite visible. Its temperature has been estimated to be about 5800K, which is not that high. But is the estimate correct?

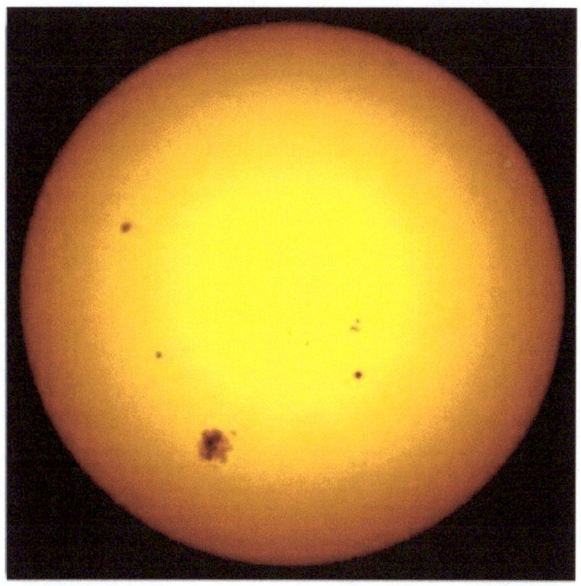

What has caught my eye is the appearance of the Sun's disc border becoming darker. Light coming from the Sun's disc border is of less intensity than the light coming from its center. At this, it is visible that the shading of the Sun's border looks the same around the equator and around the poles; that is, it does not depend on the latitude of the border. Therefore, I conclude that the actual temperature of the photosphere is the temperature of the Sun's disc border, and it is lower than the previously mentioned. It should be calculated what that temperature is and started dealing with it. When we consider the enormous gravitational force of the Sun, which creates great photo-spheric substance weight, that is, a great amount of pressure under which the photosphere substance is, then it is obvious that the photosphere is red-hot magma.

We on Earth have direct experiences with magma, which is found under the cooled Earth's crust and occasionally erupts to the surface in volcano eruptions. (Magma erupted to the Earth's surface is called lava.) The Sun's magma is much hotter than the Earth's, that is, it has a higher temperature, but it is under a larger pressure, so the word here is about a substance in a liquid state of aggregation.

Sun is, therefore, a ball made from a red-hot substance, which is very thick but in the liquid state of aggre.

Let us look again at the Sun's surface shots, without prejudice, and we will see that it is really a surface made from red-hot, thick, but liquid magma.

It is magma that is in continuous movement; warmer spurts break out to the surface, and after cooling, they sink into the depth. Logical – because the hot magma is lighter and the cold magma is heavier.

Spurts of hot magma erupting to the surface are cooled by intensive radiation and vaporization. Gas molecules, which occur by magma vaporization, have highly elevated temperatures and thus negative mass, and they are also found in the enormously strong Sun's gravitational field. What is happening there, then? The Sun is bouncing them off itself by an enormous force into the surrounding space—anti-gravitation. The repulsive force causes them to speed up, and that means an increase in their temperature, which even more, enhances the negativity of their mass, which again leads to the return of the anti-gravitational force, and so on. Due to such an abrupt rise in temperature molecules of gas are disintegrated into atoms, and then the sole atoms are disintegrated into α particles and protons. The anti-gravitational speeding process of the gas molecules from the Sun's surface is the reason for the temperature rise of up to one million degrees in the corona.

Slowly, however, owing to the huge gravitation, by the process of the antigravitational repulsiveness of the gas substance of its surface, it emits enormous energy into the surrounding space. Therefore, Sun is a much more efficient energy producer than we could have ever thought. In that way the Sun provides itself with a much, much longer lifespan than we have imagined.

Part of the electromagnetic energy that occurs in molecule and atom disintegration in the Sun's atmosphere is oriented towards the sole Sun.

heats the Sun, that is, its substance magma.

When in the initial stages the gravitational compaction heats the Sun from inside, the Sun adds the temperature to itself with the energy it creates in its atmosphere by anti-gravity.

That game of gravity and anti-gravity in the Sun and around it, finally, looks like this:

The heart of the Sun dominates anti-gravity that- balances its gravity in the outer magma layer. The Sun's at-atmosphere dominates antigravitational repulsion, which is the source of energy the Sun emits, but with the moving away from the Sun, gravity keeps the planets and everything else in rotating around it, and the Sun itself in rotation around the galaxy, dominates again.

My inner revelation that the Sun is liquid boiling magma brought me to the next realization. If we switch hydrogen, which we consider now like a fabric of the universe, with the liquid boiling lava, which is much heavier than hydrogen, it leads us to a problem that we have at that situation too much mass in the universe, and we don't see that much gravitational effects. So, it was a clear sign that stars are not full bodies of liquid boiling lava; they must have a hollow of gas inside. But I could not understand that quite well until I saw a video from the space station MIR, which shows an experiment with a one-inch rotating sphere of water with a tablet with compressed air inside. When air came to the water after the dissolving of the tablet, due to rotation, the water was rotated, and all bubbles gathered around the axis of rotation.

So that was especially important for me to see, to realize what was going on in the liquid-boiling rotating body with the bubbles inside. Inside bubbles gather around the axis of rotation because the pressure is the lowest around the axis of rotation. Centrifugal force makes that. Further from the axis, the bigger the pressure is, the more naturally bubbles follow the path from higher to lowest pressure. And that's always works. Earth's situation is that bubbles go from below to the top because they go from higher to lower, so physics is working perfectly now. Realizing that, I started to realize how Stars are born, how they develop and come to the full scale of their life. In the beginning, when the star is born from the larger star because the larger stars give birth to the smaller ones; I'll talk about that later. The first sphere is made of liquid boiling lava. It's a baby star. It is not that big in size, but due to rotation and the creation of the bubbles, because off boiling lava, some bubbles go toward the Axis of rotation, creating an inner chamber of hot gas.

During this time, that chamber will become larger and larger. Also, inflating the baby star makes the young star grow, but still, the big bubble that is inside is captured by the gravity of the liquid boiling lava around. Boiling lava is around the inner chamber of gas. It will grow until it becomes so big and due to rotation tinner in the southern and northern pole area. And then the time came for the star to start pulsating. Now we must look at this drawing of the star in their full life.

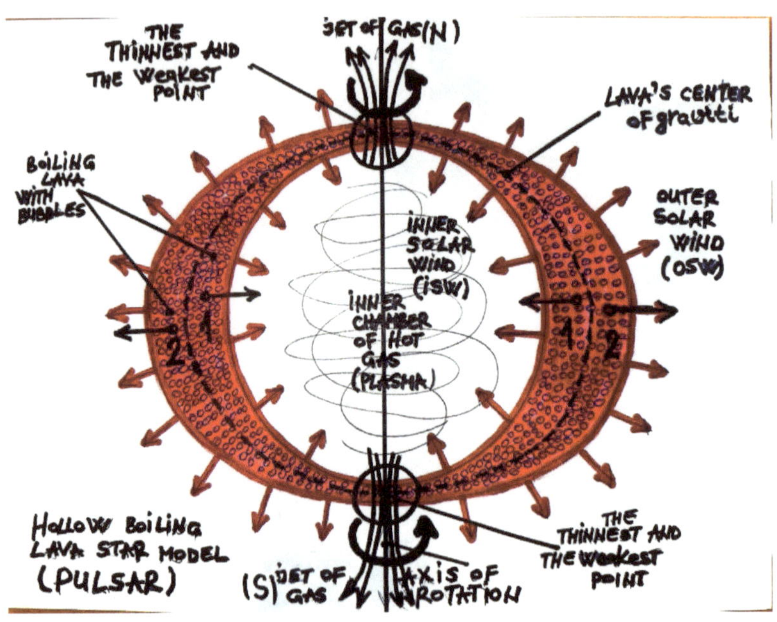
JET OF GAS (N)
THE THINNEST AND THE WEAKEST POINT
LAVA'S CENTER OF GRAVITI
BOILING LAVA WITH BUBBLES
OUTER SOLAR WIND (OSW)
INNER SOLAR WIND (ISW)
INNER CHAMBER OF HOT GAS (PLASMA)
2 1
1 2
HOLLOW BOILING LAVA STAR MODEL (PULSAR)
(S) JET OF GAS
AXIS OF ROTATION
THE THINNEST AND THE WEAKEST POINT

. . .

Now, let us discuss the drawing I made with the name Hollow Boiling Lava Star Model or Pulsar Model is what you can see on this drawing. The red area stands for the body of the star. It's liquid boiling lava. Full of bubbles, boiling means bubbles appear to the whole volume. This black line in between stands for the lava's center of gravity. This line divides the body of the star into two areas, area 1 and area 2. All bubbles from area 1 move inside the star. They create inner solar wind, and they add gas to the inner chamber of hot gas or plasma. The bubbles from area 2 go outside on the star's surface and create outer solar wind. Due to rotation, the shape is like this. Centrifuge force makes the equatorial area thicker and the polar area thinner. So, when the pressure in the inner chamber becomes so strong that the gravity of this boiling lava cannot hold it anymore, that pressure will burst through the star's body out of the star. But only through the tiniest and weakest points, which are on the star's North Pole and the star's South Pole. These areas are weakest because they are thinnest, and they have more bubbles because bubbles are gathering around the Axis of rotation. So, the bursts or jets of that inner pressure of the gas from the inner chamber will simultaneously go up and down through the North and South poles of the star. And we will see the jets of gas. We see them, and we will show you the photos made by Hubble and other telescopes like the James Web telescope, which shows exactly what I am talking about. So, the star will start to pulsate whenever the inner chamber is filled with gas with such high pressure that gravity cannot hold it anymore. It will burst through the South and North Pole of the star simultaneously, and that will give the personal life of the star. So, every star is a pulsar. Only the size of the star decides how low or high the frequency of the pulsation will be. And when we have this in mind, that stars are hollow pulsars, then there is not too much matter in the universe. It is exactly what we can see and what we can measure.

. . .

Discussing how stars create energy, I concluded that the inner chamber of higher or extremely high pressure and pulsations warms the star from the inside. They are warming the liquid boiling lava from inside the star. And there is the inner and outer solar wind which warms up the body of the star from outside. So, we have a new way of energy creation where anti-gravity force plays a key role; we can call this an anti-gravity model of the Sun or star if you like. In this way, stars create energy much more efficiently than we discussed, meaning they can live much longer than we assumed. It gives a whole new perspective on the timeline and the universe's life. With this discovery, we do not need any more Dark matter and dark energy. We do not need the Big Bang because this is an impossible thing. And, of course, we do not need black holes because they are also theoretical things made from imperfect theory where anti-gravity force was missing.

MR. THIERY'S QUESTION

Mr. Thiery's questions. After the presentation at Toronto University in July 2013, hosted by ThatChannel.TV, YouTube, and T.V. channels in Toronto owned by my friend Hugh Riley; I was sending that video to many professors from different universities and many people in science. I was getting various kinds of reactions and some questions. But the most interesting question was from Mr. Thiery, the editor of a scientific magazine in Belgium. The first question from Mr. Terry was about this picture, asking me to explain the big bubbles of hot gas above and below the center of our Milky Way Galaxy.

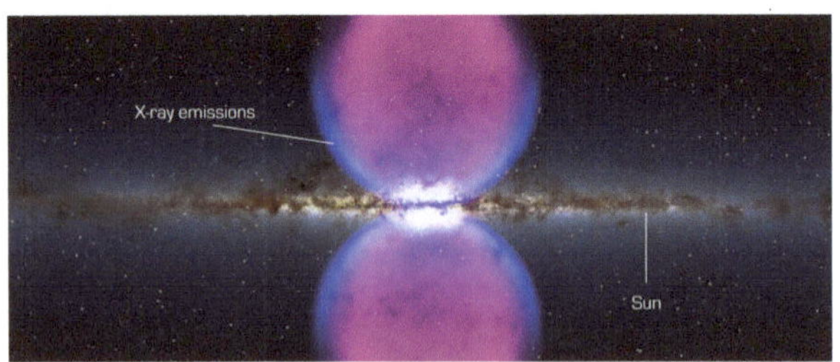

So, that was a pretty easy question for me. Because the center of our Milky Way Galaxy is the mother star of the Milky Way. That star is also a pulsar, and it works like every pulsar, blowing the gas jets from the inner chamber and making the bubbles above the North and South Pole of the mother star of our Galaxy. He was pleased with the answer. But he sent me another photo.

You can see this one asking me to explain these big rings, especially the small bright ones around the star. My answer was going like this. Big rings remain of the big pulsation. You know, an explosion on the star's North and South Pole. The bubble was created of liquid boiling lava, going away with jets, and it burst. You know, at a certain point, making these two rings made of boiling lava. But answering the question about a small bright ring around the star was a tough thing to explain, so I started to imagine everything going on there, and then the realization came like this. Bursting to the North and South Pole simultaneously, creating the two big rings, which we can see in the photo.

Two tsunamis were created simultaneously. One was on the North Pole, and the other was on the South Pole of the star, so these two tsunamis started to run and fly to the equator simultaneously. Having this simultaneous start, they splash on the equator simultaneously, creating a small bright ring around the star. Now, it stays like this, but time will show whether it will stay like this fall back to the star or fly away from it. Mr. Tiery was incredibly pleased with my answer; he said that my answer satisfied his curiosity, my logic was satisfying, and he did not have further questions for me. Thank you, Mr. Tiery,

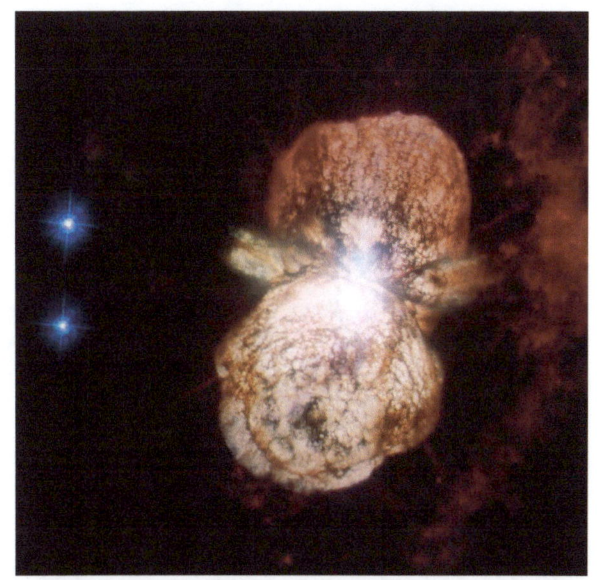

Antigravity Song and Challenging Videos

Being in Toronto at that time, I realized by following the situation about astrophysics that I needed to have a song to promote my new paradigm of the universe. With time, I met songwriter and composer Sarah Calvert, a Toronto artist and musician, and by her grace, I got the song from her. With the name anti-gravity song. I gave her words that must be in the lyrics. She wrote lyrics, and she made the song. She was performing my presentation, so you can find it and listen to it on YouTube; the link is below. Then, at the end of 2013, I was in That Channel. TV studio, with my friend Hugh Raley, I made seven challenging videos challenging two institutions. One was NASA. The other was the Canadian Institute of Theoretical Astrophysics, challenging five individuals: Michio Kaku, Neil DeGrasse Tyson, Neil Turok, Stephen Hawking, and Lawrence Kraus. You can find their videos on YouTube as well. At the time, I was ready to fight with them in discussions in all kinds of debates. But today, I just want to embrace them and hug them. Unfortunately, Stephen Hawking passed away.

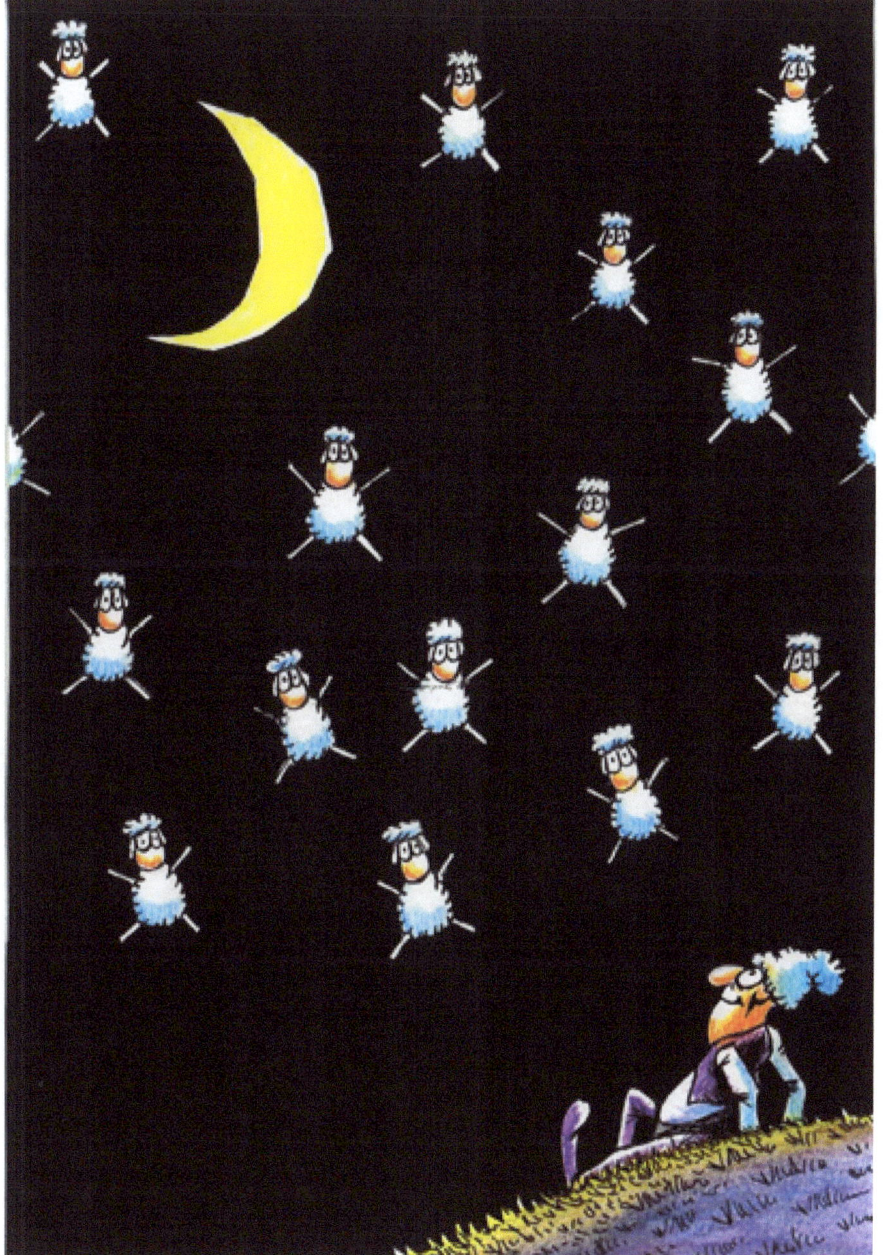

A New View Of The Sun

If now, from this perspective, we take another look at the Sun's surface and its atmosphere, we will see a completely different picture from the one that has been created up to now.

The Sun is a liquid, thick, red-hot sphere whose radius is R_0□(696000±100) km, approximately 109 times larger than the Earth's radius. The Sun's volume is 1.3 million times larger than the Earth's.

The problem in deciding the Sun's exact radius appears because of its periodic and non-periodical changes in different time intervals. The most critical short-periodical variations R are the consequence of several ways of oscillating, present in the inside and the photosphere of the Sun. Oscillations on the Sun are not only local and sporadic, but they spread through its interior, similar to the Earth's seismic waves. Because of these waves, the Sun vibrates similarly to a gong, experimentally proven in 1975. Because of that, its surface periodically, in different frequencies, rises and lowers even up to 10 kilometers (picture 3), even though the amplitudes of the global oscillations are considerably lower, and they are about 25m.

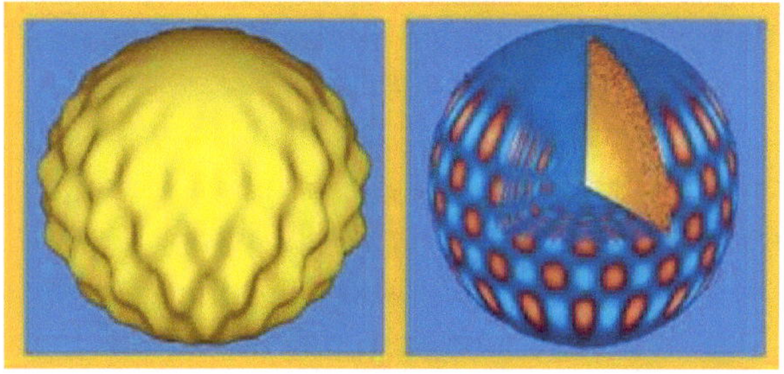

Picture 3.

Today, solar seismology is developing as a separate area of astrophysics (helioseismology). This field studies the structure, constitution, and dynamics of the Sun's interior by analyzing oscillations detected on its surface. The research method in Helio seismography is based on the analogy with the study of seismic waves on the Earth.

In the mid-eighties of the XX century, it was set up that seismic waves existed on other stars as well. Many surface characteristics of the Sun (glow, movement of the spectral line, and so on) are conditioned by the wave processes in its interior. By detailed studying and precise measurement of the wave manifestations in the surface layers, we can get information about the Sun's interior. However, we should bear in mind that the changes in the glow and radius are evoked by the small waves of the sun, and they do not exceed 0.002% of the average value.

Knowing the speed of the acoustic waves gives qualitative data about the structure of the surroundings they go through. Studies have shown that these waves do not only go through the center of the Sun, which can be understood as a confirmation of my opinion that there is a hollow in the Sun's center.

The size of the Sun has practically remained constant in the past 250 years since it has been systematically followed. However, some authors, based on observations during the eclipse, claim that the Sun's angular diameter annually reduces by about 0.0015 angular seconds. I consider this a consequence of the Earth's continual distancing from the Sun, but I will write about that later.

Galileo was the one who spotted the Sun's spinning around its axis in 1610, by following the movement of the spots on the Sun's disc from the East to the West. Based on the movement of the visible details (spots, fibers, etc.) on the Sun's disc, even in the XIX century, it was set up that the Sun spins around its axis, which makes an angle of 7,2° with the normal on the ecliptic. Rotation is happening in the direct course, which is characteristic of all planets of our system. On average, each rotation lasts about 27 days. The Sun belongs to the group of stars which rotate slowly.

In the XIX century, an especially important characteristic of the Sun's rotation was set up – it is **a differential** (zone). So, various parts of the Sun's surface rotate at different speeds (picture 4). That was irrefutable proof that the Sun is not a solid body.

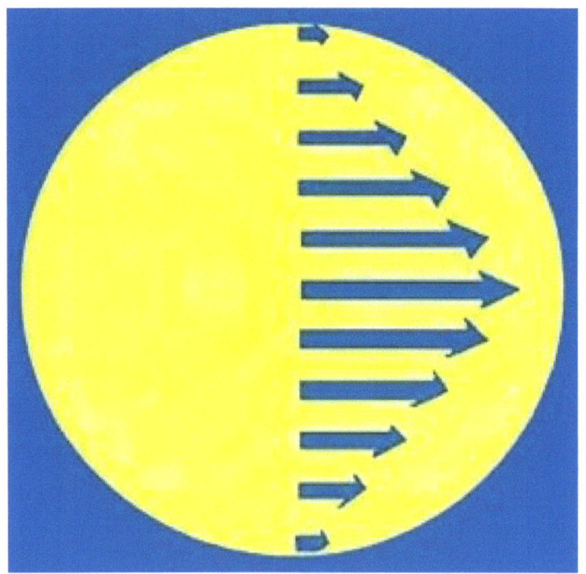

Picture 4.

The period of rotation for the points near the equator is about 25 days (peripheral speed of 2 km/s), and in the areas that are about

60o of heliographic latitude the rotation period is about 30 days. So, the speed decreases from the equator to the poles.

Oscillatory changes of the speed with time have been no-ticked as well, which can be from 10 to 20% compared to the average values. With the reduction of the Sun's activity, a slight tendency to increase the speed of differential rotation is noticed.

Jupiter and Saturn have differential rotation, at which the angular speed decreases from the equator towards the poles. On Earth, analog occurrences are spotted in the atmosphere and in the ocean.

As soon as the scientists determined the Earth's distance from the Sun a=149,6 ·109m and the speed by which from the perspective of the temperature relativity of the mass, this result is only interpretable as the total Sun's substance, according to its heat, that is, temperature and geometric allocation, is equivalent to the calculated value. I have already said that the Sun's substance is in the state of red-hot, which is a liquid magma.

At that point, of course, the density of the Sun is much greater than the density of the Earth because the Sun's gravitational force is much larger than the Earth's. By its chemical composition, the Sun is formed of heavier elements that make magma.

In addition to that, I should mention that up to today, there have been 72 elements detected by the absorption lines of the solar spectrum. That does not mean that there are not the remaining 20 elements that appear in nature on the Sun – they have simply not been detected yet.

The Sun is a ball of red-hot magma at the temperature of several thousand K, which, thanks to the antigravitational repulsion of the evaporated gas matter produces enormous energy that it emits in the surrounding space. Value of the that is a value measured by instruments and, lately, by using artificial Earth satellites. The more precise measure of its value pointed to its short-term variations with the amplitude 0,1– 0,2% from the mentioned value. The rhythm of those variations is by the solar activity, that is, the 11-year cycle of the Sun's activity. According to some research, in the past two hundred years, the mid value of the solar constant has risen between 0,25 and 0,6%.

Here, I would like to say that both the Earth and the Sun, as well as the whole Universe, are heating up and, therefore, expanding.

Apart from the electromagnetic radiation from the Sun, there are also electric particles (protons mainly) permanently going out into interplanetary space and that is the Solar wind. The Solar wind is the outcome of the decomposition of molecules and atoms' gases that the Sun is speeding up by the antigravitational repulsion and heats up to 1-2 million K in its corona.

However, the Sun is losing its substance, but very slowly, and it provides it with a much longer life than 10 billion. That is a value measured by instruments and, lately, by using artificial Earth satellites. The more precise measure of its value pointed to its short-term variations with the amplitude 0,1– 0,2% from the mentioned value. The rhythm of those variations is by the solar activity, that is, the 11-year cycle of the Sun's activity. According to some research, in the past two hundred years, the mid value of the solar constant has risen between 0,25 and 0,6%.

Here, I would like to say that both the Earth and the Sun, as well as the whole Universe, are heating up and, therefore, expanding.

Apart from the electromagnetic radiation from the Sun, there are also electric particles (protons mainly) permanently going out into interplanetary space and that is the Solar wind. The Solar wind is the outcome of the decomposition of molecules and atoms' gases that the Sun is speeding up by the antigravitational repulsion and heats up to 1-2 million K in its corona.

However, the Sun is losing its substance, but very slowly, and it provides it with a much longer life than 10 billion years, which was how much was considered up to now. Since this casts a brand-new light on the process of birth, life, and death of a star, I will pay adequate attention to that later when I speak about the stars' evolution.

Even at the end of the XVI and beginning of the XVII century, Clavius and Kepler's observations and works concluded that the Sun has an atmosphere. In contemporary times, thanks to the analysis of the Sun's electromagnetic radiation, it has become known that its atmosphere is of a layer build. As with most of the stars, in the Sun's atmosphere, three major layers can be isolated: photosphere, chromosphere, and corona.

The sun's interior is surrounded by a surface layer called the photosphere. Above it, there is the chromosphere, which can reach up to 10,000 km above the photosphere, and the corona attaches itself to it.

The average value of the layer's basic parameters in the Sun's atmosphere is said in the table below:

LAYER	Internal radius (km)	T (K)	Density (kg/m³)
Photosphere	696.000	5.800	$2 \cdot 10^{-4}$
Chromosphere	696.500	4.500	$5 \cdot 10^{-6}$
Transitional layer	698.000	8.000	$2 \cdot 10^{-10}$
Corona	706.000	1.000.000	10^{-13}
Solar wind	10.000.000	2.000.000	10^{-23}

The chromosphere and corona glow more than the photosphere so that they can be directly seen in certain situations (solar eclipse) or by special devices. During the eclipse, just a few seconds before the showing of the Sun's barrier (photosphere) behind the Moon disc, the chromosphere can be seen as a shiny dark-red line with prongs in the picture above (spicules). They can be seen not only at the Sun's border but along the whole disc and in the light of a certain wavelength, which is conducted with special devices.

The upper parts of the corona are best studied at the time of the complete solar eclipse. Otherwise, the corona gradually smears through the interplanetary space of the solar system through the Solar wind.

Starting at the bottom of the photosphere, the temperature slightly falls, to start rising above the lower chromosphere in the so-called turning layer in the chromosphere. In the transition layer between the chromosphere and the corona, temperature abruptly rises, and at certain points in the corona, it reaches the value of several million degrees. That is one reason some authors are effective between the 'cold' atmosphere, which forms the photosphere and the chromosphere, and the 'hot' atmosphere, which consists of the corona and its extension, the Solar wind.

<u>Photosphere</u>: From the Earth, it is spotted in the shape of a bright disc. It is the first transparent layer of the Sun.

Picture 5.

After the cells' eruption to the surface, the magma cools down by radiating energy and evaporating into the atmosphere. Magma thus becomes thicker, and then it sinks to deeper layers, and a new one comes into its place. In the life span of a granule, there is no mixing of magma with another granule. The approximate duration of the granule is 5-15 minutes.

It has been determined that the granules move up and down in the photosphere. By precise measuring and based on the Doppler effect, it has been calculated that they move at a speed of 0.3 to 1 km/s.

Convection in the photosphere has its own manifestation and in much larger dimensions than the granule, which is manifested in the so-called super granule (picture 6).

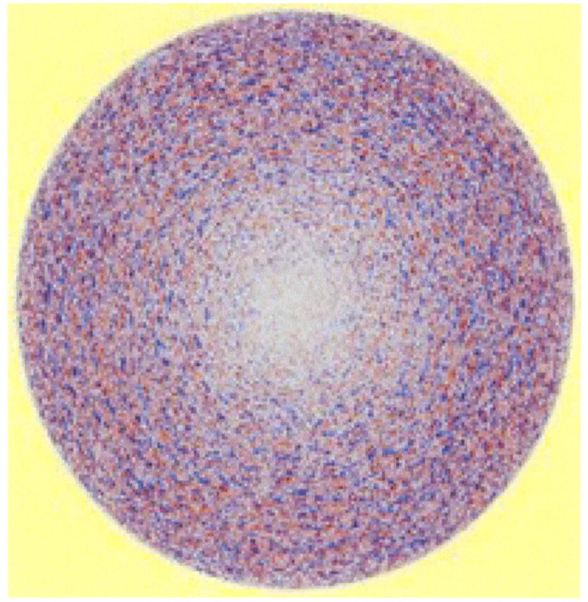

Picture 6.

They are in the shape of polygonal cells with an average diameter of about 30,000 km. They last several hours. They cover the whole Sun's surface, and their number is at each moment about 2000.

Apart from being much larger than granules, they are characterized by larger convection per the depth of the Sun's surface layer. In each super granule, symmetrically, a horizontal radial leaking of magma from the central parts of the cells towards the periphery is noted. The maximal speeds of such horizontal movements are about 0,4 km/s. In the central parts of the cells, magma from the deeper layers rises vertically towards the surface and returns to the depth along their brims. Speeds of these vertical movements are about 0,1 km/s.

In the photosphere, even larger forms of convective movement manifestations occur. Those are gigantic convective cells, and differential rotation affects their shape.

So, the Sun is, like a ball of magma, in continual turbulence and boiling. As we have seen, there are excessively big sources as well as small ones. Big sources make the basis and smaller and the smallest sources appear in them. Because of all that the Sun's surface looks like it is all in bumps that have smaller bumps on them.

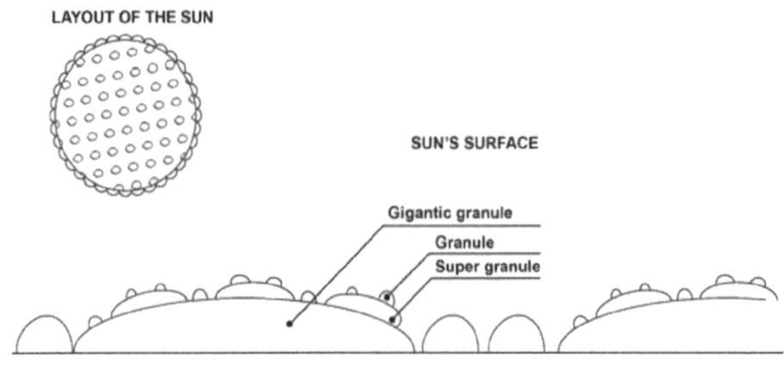

Picture 7. Granule, super granule i gigantic granule

Sunspots: Spots are one of the most important forms of photosphere activity on the solar disc. They are darker areas on the bright solar disc. For several hundreds of years, spots have been systematically checked because their number, surface they take, time, and place of their appearance provide valuable information about the activities and processes of the Sun.

Because of the enormous brightness of the photosphere and its small dimensions in relation to the Sun's disc, the spots on the Sun can rarely be seen with the naked eye, and that is in cases when they are extremely big, with a diameter over 40.000 km.

A spot appears in the shape of a dark pore, which develops afterward. They appear on the so-called royal heliographic latitudes (5° – 52°). Most often, they appear on the latitudes of 8o to 30°.

The diameter of the smallest spots is the granular dimension (about 1000 km), and the biggest, the so-called spot groups, even up to 100,000 km. smaller spots often last less than two days, and most of them disappear the same day they appear. Developed spots last 10 to 20 days, and the largest ones up to 100 days.

At the developed spot, a darker shadow and a lighter semi-shadow can be seen. Shadow and semi-shadow are visually clearly differentiated (picture 8). On average, the diameter of a shadow is about 17.500 km, and of the shadow about 37.000 km. The surface of a typical spot is about 10-thousand part of the visible surface of the Sun.

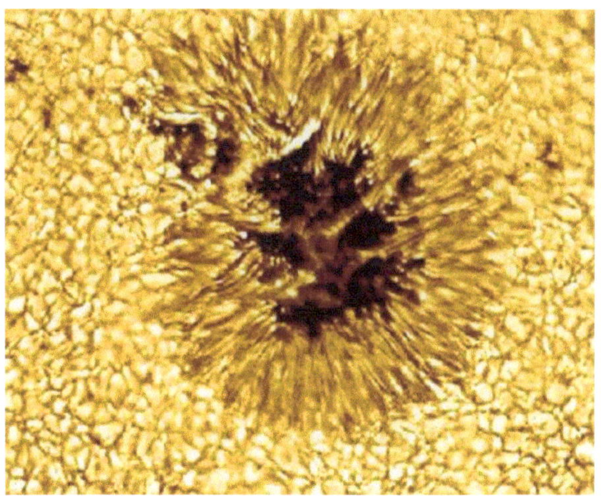

Picture 8.

The glow of the shadow is only 20–30%, and of the semi-shadow 75–80% fons of the undisturbed photosphere. A spot looks dark (and cold) compared to the high glow of the photosphere. Regardless of that, the glow of a spot of an average size is about 5000 times larger than the glow of the Moon. The lowered level of the glow in the spot area is compensated by the rise of the glow around it, at a distance of 50,000 km from its center. The glow is about 3% higher than the average photosphere glow.

The temperature in the spots is 25–30 % lower than the photosphere. The temperature of a shadow decreases with the growth of its surface.

Granulation is present with the spots, too, but in a different form compared to the photosphere, which is a consequence of the changed convection. Granules in the semi-shadow have the shape of light fibers about 300 km wide. Fibers last about from 30 minutes up to several hours, which is much longer than the durability of the granule in the undisturbed atmosphere. Granules in the shape of bright spots can be seen in the shadow, too. Their duration is about 15 to 30 minutes, and their dimensions are about 350 km.

It has been noted that with the spots nearing the west border of the Sun's disc, the east half of the semi-shadow gradually begins narrowing and disappearing. Upon the appearance on the east border of the disc, the semi-shadow is not visible at the beginning, and it appears, and it starts becoming wider, but first, it's the west half. A spot is a shallow funnel-like cavity in the photosphere.

Sometimes it can be noticed, and it is not so intensive a rotation movement, whose speed at the end of the semi-shadow reaches 14km/s. With the solitary spots, this rotation can lead to the formation of the vertical structure of the elongated details of the spot.

It is especially important to point out the existence of the warmer areas called photosphere faculas (torches). They are long-living bright areas that, in principle, do not have to be connected to the spots (picture 9).

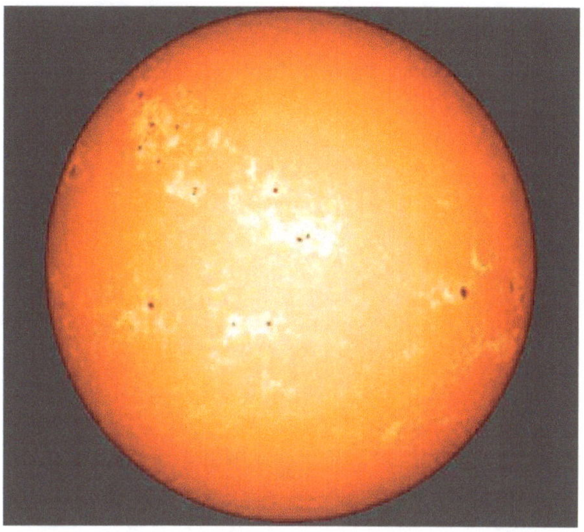

Picture 9.

In the photosphere, there are independent faculas, as well, that are of a lesser brightness than those surrounding the spots. I especially enhance that there are no spots without faculas. They have a granulated structure that is 10% brighter than the surroundings. Inside them, granules last longer (about an hour) compared to the granules from the rest of the photosphere. The dimensions of the granules in faculas are about 1000 km, but they form groups 4-6 thousand km long. They connect in chains 5 to 10 thousand km wide and up to 50 thousand km long. Large faculas appear several hours or days before the spot, and they are visible long after the spot disappears. Often, they do not disappear even for a year. On average, they 'live' twice as double the spot they form around and occupy about four times larger surface than the spot.

Spots, therefore, always appear in faculas and as solitary, or, more often, they appear in groups (picture 10).

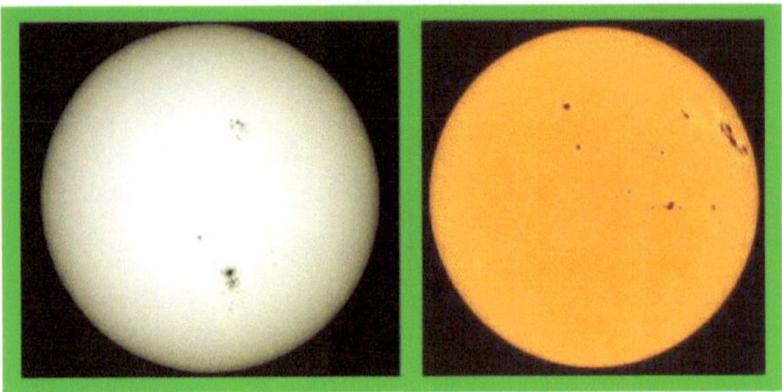

Picture 10.

Solitary spots are often the beginning of a reminder of a larger group. The number of spots in a group can be up to several hundred. Large groups of spots reach their maximal development in two or three weeks, and then they slowly fall apart and vanish during 1,5 up to 2 months. When a group reaches its maximum (according to the number of spots, surface it takes, etc.), it starts receding spots fall apart, and their number in the group diminishes. The process of the spots falling apart usually happens in such a manner that brighter spots appear over them.

On average, groups of spots last 10 days (30% of them last more than 10 days, 0.4% more than 50 days, 0.3% more than 100 days, and only 0.01% last more than 150 days). The larger the surface a group of spots occupies, the longer its duration. For example, if a group of spots occupies the 10-thousand part of the visible semi-sphere of the Sun, then on average, it lasts 10, and if it occupies 4 times larger the surface, it lasts 40 days.

Spots and their groups can have different shapes; sometimes their shape is regular and oval, and very often it is irregular (picture 11).

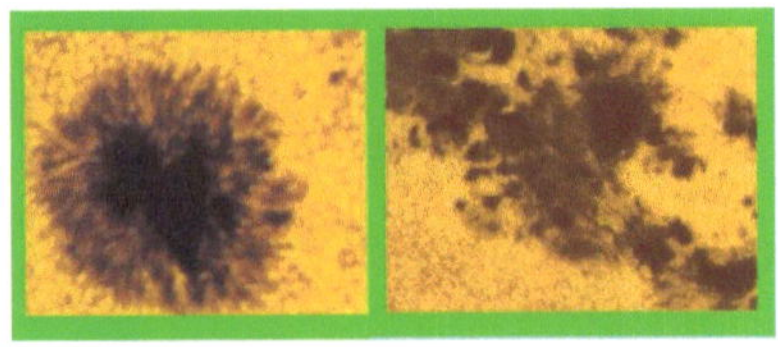

Picture 11.

If it concerns a couple or a group of spots, then the west spot is the leader, and the other spots are followers. The leader spot is the one that appears first in the couple or a group of spots.

The leader is closer to the equator; on average, it is larger than its followers and lives longer. In the beginning, the leader moves faster than the followers, and the formed spot group is dragging. When the furthering of the leader in the group stops, the process of destruction and the disappearance of the group begins.

Spots can be without semi-shadows, and they can have smaller spots in their semi-shadows. By falling apart of the big spots, small and middle-sized groups occur.

When we have said and conjured up all this about sports, it is time for revelation. Simplicity in all its perfection.

We have seen that the surface of the Sun consists of granules, super granules, and gigantic granules. Granular is the geyser of magma. Therefore, the Sun's surface is all in geysers of magma; basically, those are gigantic geysers of magma, and the super-geysers of magma exist, and on all of them, small geysers of magma are present. What is mutual for all these geysers is that they eject magma to the surface from a relatively thin and homogenous layer which is the outer layer of the Sun. That is a game of boiling of that outer layer, and that is what is the normal state of the Sun's surface.

What is unusual is the appearance of chromosphere faculas or torches. Faculas are the Sun's volcanoes. Faculas are processes of ejecting magma from the larger depths, that is, from the Sun's deeper layers to the surface. That is why the faculae are brighter and warmer than the rest of the photosphere. We have seen that those volcanic eruptions can eject magma of different temperatures, which means from different depths. These volcanic solar eruptions can be realized from the beginning to the end in that manner.

But when a large solar eruption is in question, it not only ejects magma from the deepest surface layer of the Sun but also starts ejecting qualitatively different magma that is located below that surface layer. That magma has a different chemical composition compared to the surface magma, which consists of **heavier chemical elements**. When this magma is ejected to the surface, it cools down quicker, gets thick, and starts floating along the Sun's surface made from geysers. We see that heavy magma, from the great depth of the Sun, we see as a black spot. Since it is heavier from the surface magma, it makes a cavity in the photosphere, and since on its brims it is thinner than in its central part, the ends are easier, and they form slopes to the level of the photosphere, and that is how the funnel appearance of the spot is achieved. The size of the spot, as well as it looks, is explicitly defined by the quantity of the heavy magma ejected in the eruption of the Sun's volcano (facula). The number of spots is a consequence of the intermittent ejecting of the heavy magma to the surface. The first amount of heavy magma is the leader spot, and the rest of the heavy magma forms the follower spots. The first amount is the largest because it is the greatest relaxation of the intensity that led to the eruption, and that is why the leader spot lasts the longest. If the ejecting of the heavy magma is continual, then we see the spot widening until there is a break in the eruption of the magma.

Differential rotation of the Sun carries the spot towards the west, and the continuance of the eruption creates the next spot, and so on until the eruption of the hot magma stops. That is how spots appear. Cooler, heavy magma floats on the geysers of the warmer surface magma, which will slowly but certainly 'melt' it. Magma geyser pressure to the spot, that is, heavy magma, from above leads to the perforation or the breakage of the spot into parts, which accelerates the disintegration of the spot. It is a process of reducing, falling apart, and disappearing from the spots. Sun's volcano (facula) will continue ejecting warmer easy lava, and then it will stop doing that, too. Then the facula disappears from the photosphere, that is, the Sun's volcano is extinct.

When we have realized what happens on the Sun's surface, we can go farther into the height of the Sun's atmosphere. First, let's see what contemporary astrophysics has to say; what have we discovered and seen so far.

Above the photosphere, there is the chromosphere (picture 12).

Picture 12.

The radiation of this layer is the most intensive in the red part of the spectrum, so it got its name from its intense color. The chromosphere is not homogenous, and it is quite roughly divided into the lower (1500 km above the photosphere), middle (between 1500 and 4000 km), and upper (from 4000 to 10,000 km). The lower chromosphere is relatively homogenous, and there, going from the photosphere, the temperature continues falling to the values that are lowest on the Sun and which are about 4200 K in certain places. Below the lower photosphere, in the rotating layer, the temperature starts rising, and by its peak, it reaches 10.000K.

With its height, the level of ionization grows. The upper chromosphere is very ionized, which is quite understandable when it is known that the temperature reaches values of 25.000 K there (in some places up to 300.000 K).

The chromosphere is also characterized by intensive turbulent movements. At a height of 500 km, the speed of the turbulent movement is about 5km/s, and the speed of the turbulent movement at 5000 km is about 20km/s.

Based on the monochromatic shots of the chromosphere, it is noted that it, similarly to the photosphere, has a grain-like structure. Grains are in the shape of fibers and flocculate. They are larger than the photosphere granules and can be several thousands of kilometers long.

Flocculates are well spotted on the large bright surfaces and chromosphere torches (chromosphere facula) (picture 13).

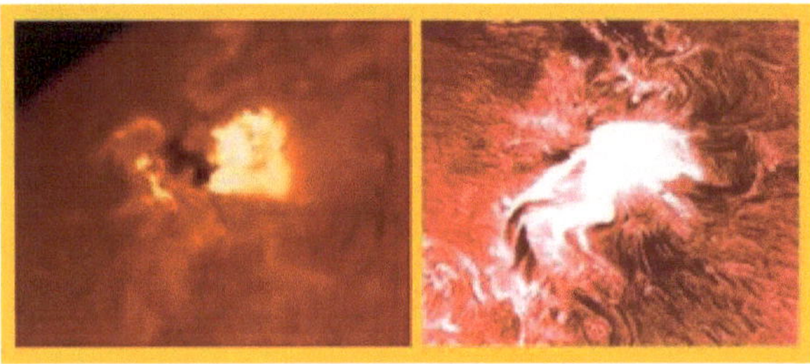

Picture 13.

When the photosphere is seen in white light, we see that photosphere faculae are in the same places as the chromosphere torchers. The word is about the same objects, seen from different heights. Chromosphere facula last longer—200 to 300 days—and they are much brighter than the font (in the X-ray area, even up to 70 times).

One of the most important shapes that appear in the chromosphere is the chromosphere network (picture 14).

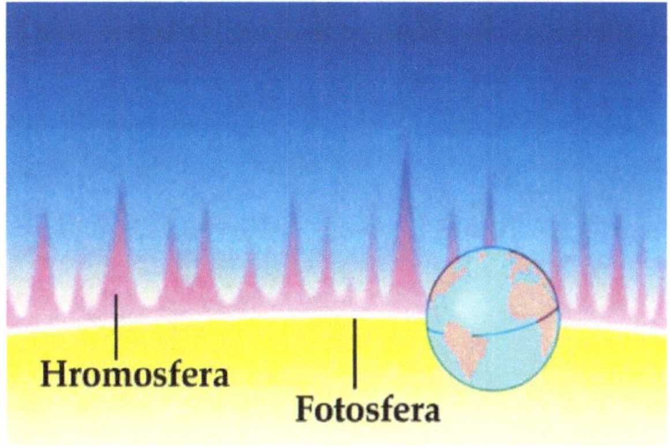

It is the result of the action of the super granules that is formed in the photosphere, but their influence is displayed in the chromosphere, too.

The chromosphere network spreads vertically along the whole chromosphere height. The dimensions of the network structures agree with the dimensions of the supergranule.

Along the brim of the super granule from the lower layers of the chromosphere, like fire tongues, spiculae go up. Spiculae represent small eruptions of hot gas, whose temperature is about 15.000 K. Particle concentration in them is 10^{18} m^{-3}, they appear on heights from 3000 to 4000 km, and they go up from 7000 to 12000 km. Their diameter is the size of the granule, about 1000 km, and they appear above the boundaries of the super granule. It is them that represent the thin structure of the chromosphere network, which is noticed in the center of the Sun's disc. It is estimated that at each moment, there are about a million spiculae in the chromosphere. They are not equally distributed. In the polar areas, there are about 30% more than in the equatorial area. They take up only 1% of the total surface of the Sun. Gas in the spiculae rises from the lower layers of the chromosphere at a speed of 20km/s.

One of the more important manifestations of the Sun's activities is eruptions (explosions) in the chromosphere. They are sudden, short processes that result in a large increase in the intensity of radiation in the limited areas of the chromosphere.

Explosions appear in the areas of the facula, above the groups of spots. They are characterized by the rapid rise in glow, noticeably short duration of their maximum, and relatively short extinction, which lasts about two times longer than the period of glow rise. When the explosions are noted at the brim of the Sun's disc, we can see a cone of light, whose height is several thousands of kilometers.

Smaller explosions, which are met more often, have a circular form, while the larger ones have an extended shape and fiber structure. Such explosions occur rarely and only at the time of the Sun's maximal activity.

At the time of the Sun's maximal activity, energy freed on explosions in the higher layers of the Sun's atmosphere is comparable with the energy it radiates in one second.

Many particles move at the speed of 1500 km/s (with extraordinarily strong explosions up to 2400 km/s) through the corona and interplanetary space (picture 15).

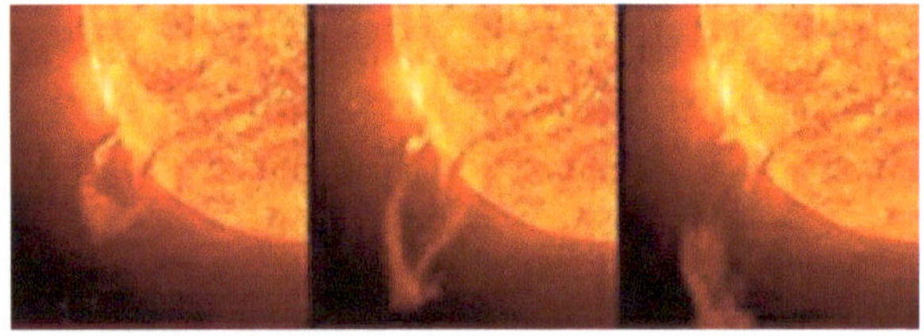

Picture 15.

Some of the particles speed up to relativistic speed, so they get to Earth almost at the same time as the electromagnetic radiation emitted in the explosion. Bunches of such particles are known as the Sun's cosmic rays.

Smaller explosions last from 5 to 40 minutes. On average, every seven hours (at the time of the maximal solar activity, every two hours) in the life of a group of spots, there is one chromosphere explosion. During the transit of a spot group through the Sun's disc, 30 to 50, and at the time of the Sun's most intensive activities, up to 300 chromospheric explosions appear above it. On the whole Sun, daily, about 100 ex-plosions occur, but those considered strong ones are very rare, and they happen only a few times a year.

During the explosions local heating of the substance occurs up to temperatures from 10 to 1000 million K.

Explosions in the chromosphere are closely related to the averages in the corona and its characteristics. This is confirmed by the fact that the so-called corona's condensation appears on heights of 10 – 40 thousand km above the chro-atmospheric explosions. Radio flashes from the corona testify about that, and the radio flashes are, on the other side, caused by the level of the activities in the photos here,

And now, the real explanation of the occurrences in the chromosphere.

The red-hot magma erupts to the Sun's surface in geysers and is cooled down by intensive radiation and vaporisation. In the beginning, in the lower chromosphere, a fall in the temperature occurs compared to the photosphere. But the The process of the gases cooling down is dominant only up to 1500 km, while the speeds of the gases' rising are small, about 5 km/s. That is, in fact, the very beginning of the anti-gravitational bouncing of molecules of the evaporated and ejected gas from the red-hot magma.

As soon as the gas molecule's speed rises to 20 km/s, at the height of about 5000 km, a rise of the temperature occurs again. The rise in the gas molecule's temperature leads to the rise in its negative mass, which leads to the rise in the anti-gravitation force by which the Sun bounces it off itself. It normally leads to an even larger speed of the molecules, even larger heating, and thermic ionization, which, as we have seen, rises with height.

Since granules, or grains, represent geysers of magma, where the evaporation is the most dominant, that structure is transferred upward, and that is why the chromosphere looks as it looks in fibers or focus.

Chromosphere torches occur above the photosphere facula, that is, the Sun's volcanoes. Logically, the enhanced temperature of the photosphere magma geysers provokes enhanced evaporation with a higher gas molecule's start temperature; thus, the antigravitational bouncing is much stronger, which leads to much larger speeds and much higher temper features than in the surrounding chromosphere.

The reticular look of the chromosphere and spiculae is just a logical consequence of the look and the activities on the Sun's surface.

Eruptions or explosions in the chromosphere occur above groups of spots. Logically, groups of spots are created by the strongest Sun's volcanoes, and they, apart from ejecting flares of 'chemically heavy' magma, eject flares of hot gases, too. These flares of hot gases from the Sun's interior have a much higher temperature than the vaporized gases from the surface magma that is cooling down. That is why they are affected by a much higher antigravitational Sun's force, but the anti-gravitational force of the mutual bouncing of the gas molecules, as well, so that their much more intensive speeding now looks like an explosion. Here, as we have seen, the particles move at the speed of 1500km/s up to 2400km/s. Some of the particles in that process carry out speeds close to the speeds of light, the so-called relativistic speed. And that is logical when we can see that the local heating goes up to 10 or 100 thousand $^{\circ}$K.

Explosions in the chromosphere are further reflected in the looks of the Sun's corona. So, everything begins with activities on the Sun's surface, and it

is only adequately separated from the higher layers of the Sun.

<u>Corona</u>: above the chro- atmosphere, above the thin transitional layer, there is a corona (picture 16).

That is the hottest and the scantiest layer of the Sun's atmosphere were.

Picture 16.

It is noticed as pearly- silverish light that encircles the dark Sun's disc at the time of the Sun's eclipse. It is much paler than the chromosphere, nonetheless, it is more visible than it since it is the most spacious layer of the Sun's atmosphere.

Dimensions and the shape of the corona, as well as the processes taking place in it, on a larger scale, depend on the activities of the Sun. At the time of the minimal activity, it is compressed above the poles and with characteristic fan-like shapes, while along the equator, it is elongated. In the period of maximum activity, the corona takes the ' disheveled' shape and almost symmetrically encircles the whole Sun (picture 17).

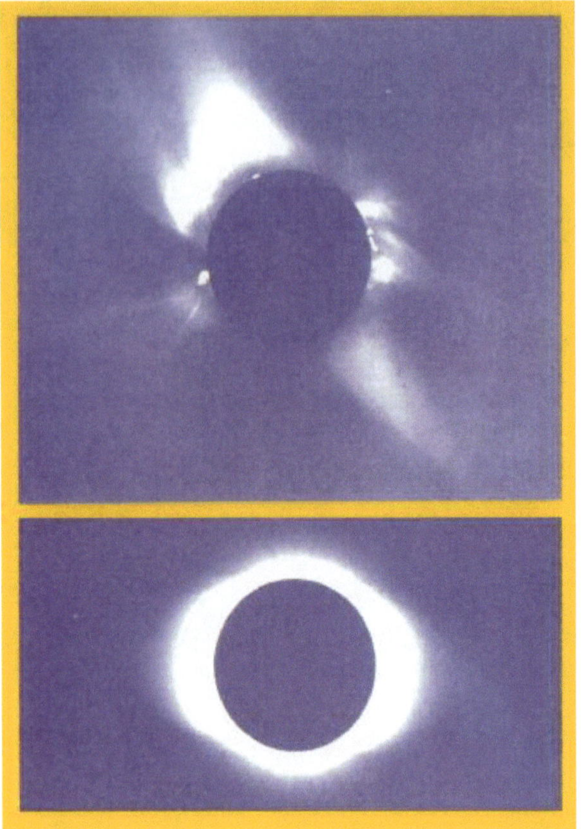

Picture 17.

Corona extends up to several Sun's radiuses, although the upper border cannot exactly be decided; through the Sun's wind, it gradually crosses to interplanetary space. Many authors treat the Sun's wind as part of the Sun's corona.

The corona is separated from the chromosphere by the al-already mentioned transitory layer in which temperature sharply rises from the chromosphere ($\geq 10^4$K) to the coronal ($\geq 10^6$K) values.

Corona's temperature reaches a maximum of about.

2 million at the height of about 1/10 Sun's radius of the photos-

phere. After that, it gradually falls. If the Sun's wind is treated like an extended part of the corona, then in the vicinity of the Earth's orbit, its temperature is of the measurement order of 10^5 K.

The physical state of the gas in the corona is defined by the surprisingly elevated temperatures and very low densities of gas.

In the inner corona, r = (1,03 – 1,2) R☉, going towards the periphery, the gas density reduces, but the temperature rises, and at the height of 50,000 km, it is (1–1,5) million K. It is characteristics that the inner corona emission spectrum with highly atomized metal atoms appears. The highly atomized metal atoms are especially bright.

In the middle corona (r ~ 2R☉) T ~ 10^6 K gas density decreases even more and its radiation is with an elevated level of In the outer corona, r ~ 3R☉, temperature, and gas concentration values continue falling, and the corona gradually starts squandering into the interplanetary space. The brightness of the outer corona is about a billion times lower than the brightness of the Sun's disc. The continuous spectrum of the outer corona is practically the repeated spectrum of the photosphere but of drastically lesser intensity. In it, we can also see standard Fraunhofer's absorption lines, which are characteristic of the photosphere. That points out that in the outer corona, dispersal of the radiation of the photosphere on the present dust particles happens.

Sun's corona is visible in the wide diapason of the spectrum of electromagnetic radiation.

The light of the emission corona consists of about 100 bright lines, which have been determined to originate from the highly ionized atoms of iron, calcium, nickel, etc.

Corona's radiation arises in conditions that are much different from the thermodynamically balancing ones. Owing to the elevated temperatures, the corona is a strong source of radio- and far UV radiation. Newer research denotes that in the corona, there are occasional flares with intensive emissions of X-ray radiation, which precede ultraviolet emissions. Radiations originate from the area in the corona where temperatures reach up to several million degrees.

In the corona, we can note different forms: flares, rays, arches, plumes that appear above the poles in the shape of brushes, condensations and hollows, eruptions, etc. Some of them are visible in the integral light of the white corona, while others are more visible in other areas of electromagnetic radiation (radio or X-ray). Denominated shapes are most often found above the area of the enhanced Sun's activity in the so-called active areas.

A flare manifests in the abrupt enhancement of the glow of the localized areas on the Sun above a group of spots and torches. This occurrence happens in the areas of the chromosphere and corona. In about ten minutes, the glow of the flare overtook the area, and then, for about an hour, it reduced and returned to the former level.

Corona's rays are characteristic elements of the larger forms in the Sun's corona. They are noticed during the solar eclipse or with the help of a chronograph as elongated condensations with different radial forms (picture 18).

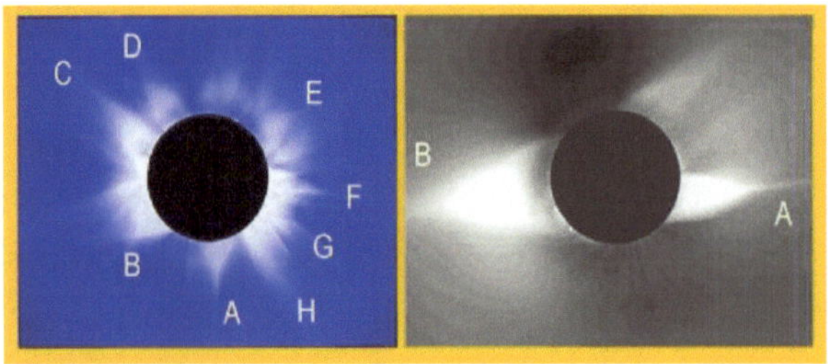

Picture 18.

They extend from 0,5 R☉ to 10 R☉, and in some cases even more. On average, the duration of the ray shapes is about ten days. The main part of the ray, above the center of the activity, emits a green line $Fe13+$ so that, in that area, the temperature is over 2 million degrees.

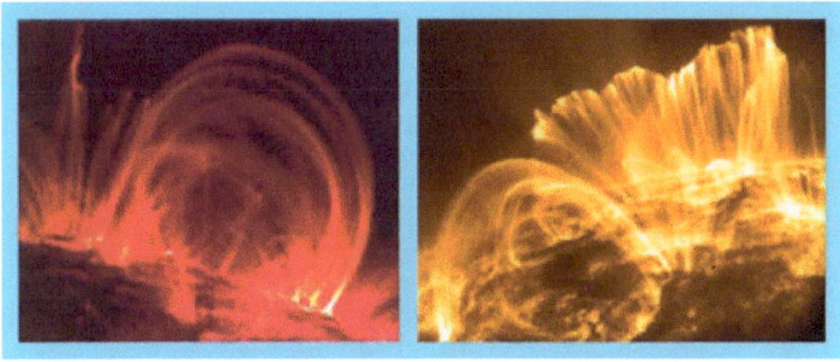

Picture 19.

Corona's arches (picture 19) occur above the field of the active areas. They are composed of a number of thin threads.

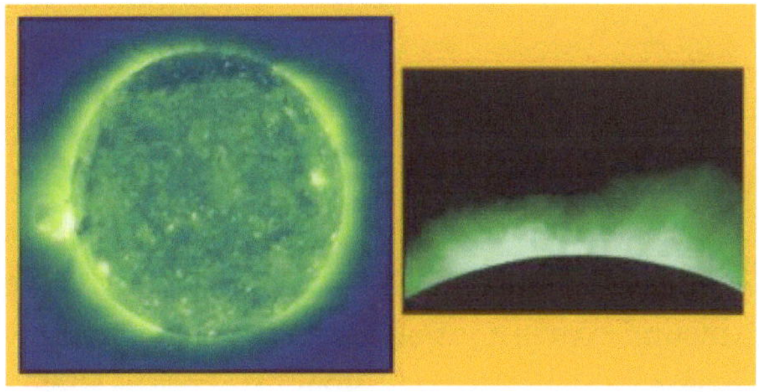

Picture 20.

Plumes or polar brushes (picture 20) most often occur above the Sun's poles. They greatly depend on the level of the activities on the Sun, they connect to the present faculty, and they are best seen during the time of the decreased activity when the corona is compressed.

Analysis of the radio waves from the Sun has shown that above the central parts of the active areas, there are corona condensations. Their temperature reaches values higher than three million degrees. They are several times thicker than the corona's surroundings. At the time of the most intensive creation of the spots, sporadic condensation occurs.

In the Sun's corona, areas with lower temperatures (from 0.8 million K) and an anomalously low mass density were spotted, and they were named <u>corona's hollows</u> (picture 21).

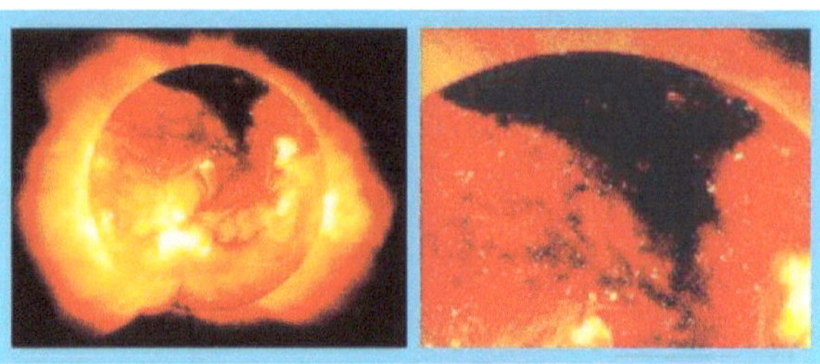

Picture 21.

The word is about spacious, stable formations, that sometimes accept up to 20% of the corona. They last about several Sun's rotations. They are characterized by the reduced glow in the X-ray and the UV area of radiation. A noticeable decrease in the glow in the radio and the visible part of the spectrum. Judging by everything, they permanently exist in the polar areas of the Sun, but they sometimes expand to the lesser heliographic latitudes. There, isolated hollows can be formed. Hollows like this are often long-lasting, especially at the time of reduction of the Sun's activity. From the corona's hol- lows, the Sun's wind, whose particles leave the corona at speeds from 600 to 800 km/s, is continuously emitted.

On the recordings from the corona in the X-ray light, the <u>Corona's bright spots</u> are noted, unorderly distributed over the whole Sun's disc. Their dimensions are smaller than the dimensions of the spots, and on average, they last for about 8h. During the 24 hours, on the Sun about 1500 corona's bright spots appear. The dynamics of their appearance and their total number are in antiphase with spots; that is, they occur more often and in larger numbers when there are fewer spots in the photosphere.

<u>Protuberances</u> are the most spectacular form of the Sun's activity and the most grandiose occurrence in the Sun's atmosphere.

During the solar eclipse, they can be seen as red flames (picture 22).

Protuberances are strap condensations in the corona. It is about cooler (T $\leq 10^4$ K) thicker formations inside *Picture 22.* diluted and warmer coronas (T ≥ 106 K).

Above the Sun's limbo, protuberances can be seen in the shape of gigantic fire tongues, arches, fountains, knots, etc. (picture 23)

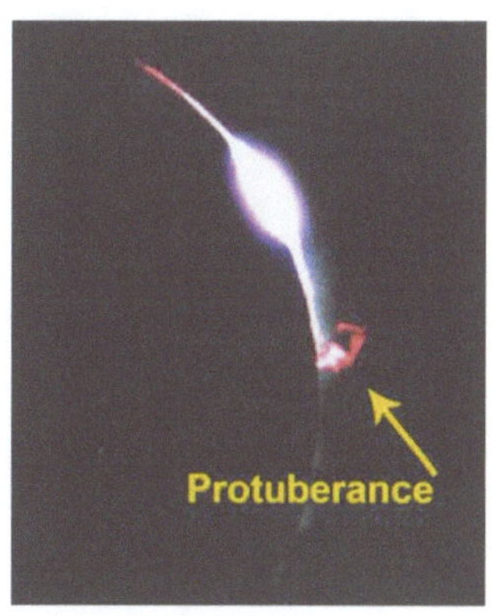

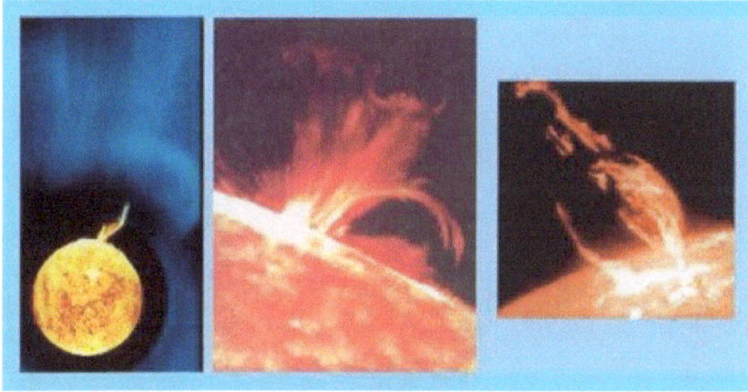

Picture 23.

We can spot moving threads and condensations in protuberances. In the projection to the Sun's disc, they are dark, bent strips of complex structure called filaments. Their length is up to 200,000 km.

The best-known distribution is to quiet and active protolerances.

Most protuberances belong to the quiet protuberances. They are 'long-living' - they last more than one day (which is rare) to several months (which is more often). There were cases that lasted for several years. They hover for a long time on all heliographic latitudes. The temperature of the quiet protuberances is usually up to 15000 K, and most usual.

between 6000 and 8000 K, which is the reason they are treated as cold ones.

A typical quiet protuberance is about 200.000 km long, even though in rare cases their length can reach a total of 1.900.000 km. The average height is 50,000 km, while the width is not more than 6000 km. They consist of threads whose diameters are about 1000 km. The lower ends of protuberances are in the areas between the supergranule and in the vicinity of the active areas.

Protuberances primarily occur in the zones of latitudes of 10°–40°, in which spots are concentrated, but they spread further. Quiet protuberances are divided into those below the 40° ⁻ 45° heliographic latitude and those above, known as the polar, which often make the so-called wreaths of polar protuberances (picture 24).

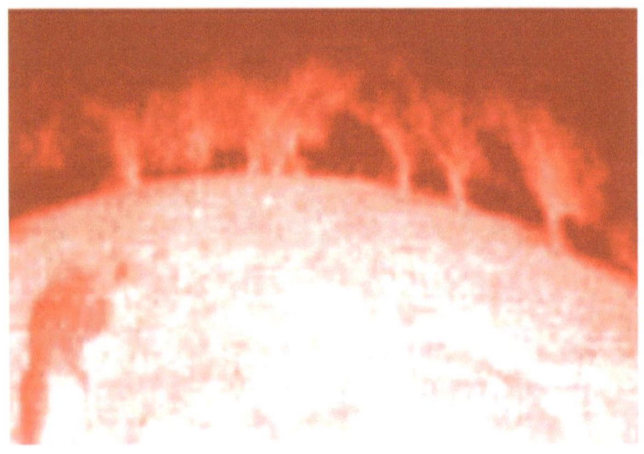

Picture 24.

In the beginning, quiet protuberances usually spread in the direction of the meridian and in such a manner that one foot of a protuberance has 80% of its fibers directed toward the leader spot. In time, they slowly move towards the pole. Their length is growing, and they are increasingly oriented east-west. After ten or so of the Sun's rotation, their threads reach the polar areas, where they can survive about five more months.

Apart from the quiet ones, there are <u>active protuberant- dances,</u> <u>which distinguish themselves by very rapid development</u> (from 10 minutes to several hours). They appear in the shape of clouds, systems of knots, cyclones, sprays, etc. They are usually of smaller dimensions than the quiet, protuberant- dances. Some of them appear in areas above spots and slowly move towards the borders of the area. They are very often absorbed by the spots, and they disappear there. They can be transformed into quiet ones. The average temperature in active protuberances is about 25.000 K, which is why they are known as hot ones.

In the spots' zone <u>eruptive (explosive) protuberances</u> appear. They reach great heights – in some cases over a million kilometers. There's a known case of a protuberance that 'rose' up to 1.700.000 km (picture 25).

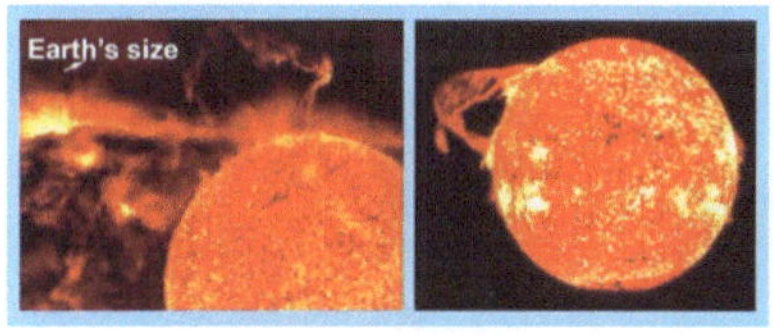

Picture 25.

In the spectra with, lines typical for quiet protuberances (hydrogen, lines of calcium, etc.) appear, but lines of metal also appear.

At the beginning of the formation, for a while, eruptive protuberances can resemble regular ones, the quite protuberant- dances that have in themselves elements with unordered movements. Parts of the protuberance then begin to rise, at first slowly and then faster and faster. The rise in the speed can be abrupt and thus reach speeds up to several hundred km/s. The mutual feature of all eruptive protuberances is that they 'break' in the middle of the arches. During the arch eruptive protuberant- dances (picture 26), it comes a great enlarging of the size of the arch. After its breaking, the matter returns to the chromosphere down along the parts of the arch.

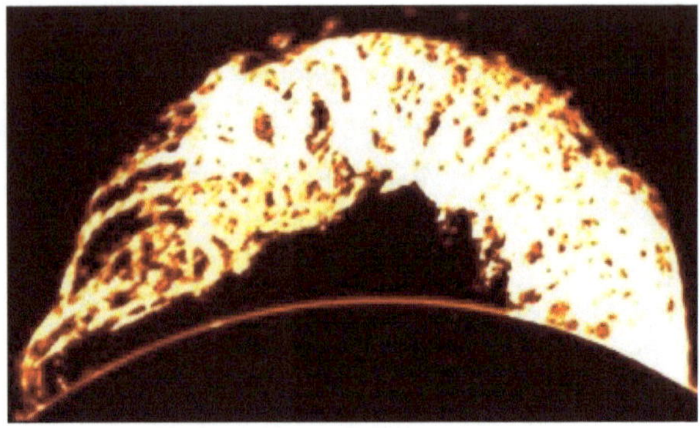

Picture 26.

There are the so-called protuberances of the sunspots (picture 27) that appear in the centers of activities above the spots' groups in the shapes of knots, arches, or funnels.

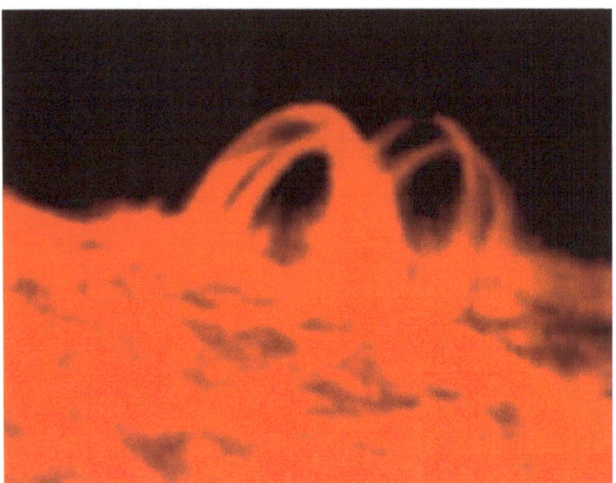

Picture 27.

There's also <u>corona's rain</u> – a form of protuberance that occurs with the fall of the tufts from the corona to the chromosphere.

Because of the greater mass density, substances in the protuberance should 'sink' to scantier environments. However, a certain force enables them to survive relatively long in the formed configuration.

<u>Sun's wind</u> occurs through the expansion of the corona into interplanetary space. It is a permanent 'circulation' of the Sun's substance that moves approximately radially from the Sun (picture 28) and pervades the Sun's system up to the reach of 100 AJ (AJ is the distance from the Sun to the Earth).

Picture 28.

Particles of the Sun's wind gradually speed up, for example, at the distance of several Sun's radiuses, the middle radial speed of the proton collective in the Sun's wind is (100–150) km/s, and at the distance of 1 AJ, its value is (300- 750) km/s (picture 29).

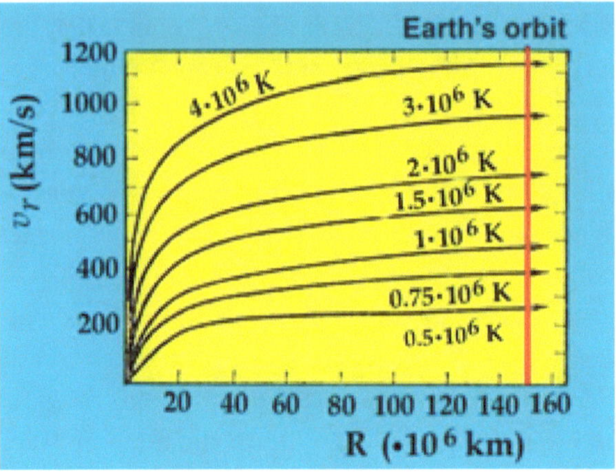

Picture 29.

Regarding the speed of movement, flows of the Sun's wind can be either slow (with speeds up to 300km/s) or fast (with which the speed has the value of 600-700 km/s).

Near Earth's orbit, depending on the level of activities on the Sun, through the square meter of the transversal surface, every second between $5\cdot10^{11}$ and $5\cdot10^{12}$ protons whose average speed is 400 km/s, and the average temperature is.

50.000 K (in active periods, it can be up to 400.000 K) 'flow'.

In a second, through the Sun's wind, into interplanetary space, a mass of 10^8–10^9 kg flows out. If there weren't for spiculas, which are 'filling' in the corona's mass, this layer of the Sun's atmosphere would be 'scattered' in 3 or 4 days because of the flow of the Sun's wind. Due to the solar wind, on the annual level, the mass loss of the Sun is 10^{15}–10^{14} M☉. Since, judging by all, the Sun's life span is about 10 billion years, it is clear that such a loss does not significantly influence its evolution.

Table 2. Relative content of atoms in the solar wind's chemical structure

Element	Relative content	Element	Rel. sadžaj
H	0,96	Ne	$7,5 \cdot 10^{-5}$
^{3}He	$1,7 \cdot 10^{-5}$	Si	$7,5 \cdot 10^{-5}$
^{4}He	0,04	Ar	$3,0 \cdot 10^{-6}$
O	$5 \cdot 10^{-4}$	Fe	$4,7 \cdot 10^{-5}$

Table 2 presents solar wind's chemical structure with the relative content of certain atoms compared to the total number. The representation ratio of helium and hydrogen atoms in the Sun's wind is not the same as in the Sun's atmosphere (to the detriment of helium). However, in the quite solar wind the representation ratio of ^{3}He: ^{4}He coincides with the same ratio in the remaining part of the solar atmosphere.

In the Sun-related coordinates system, the Sun's wind particle trajectories (paths) are in the shape of Archimedes' spirals, with the origin in the areas of the Sun's corona from which the mentioned particles originate (picture 30). The spiral line of the solar wind's 'blow' is decided by the radial speed of the particles flowing out from the corona.

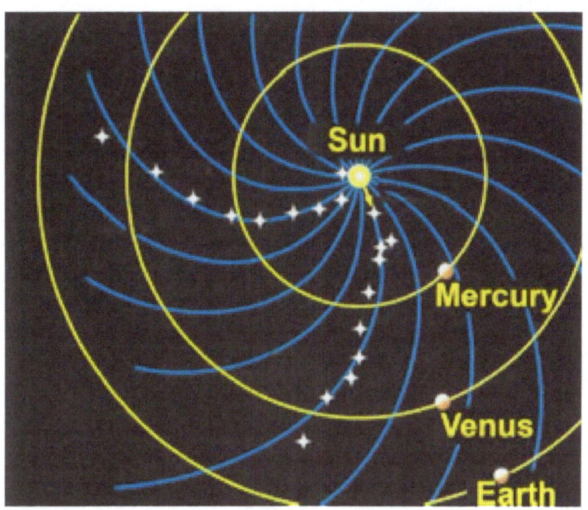

Picture 30.

The area of the Sun's wind and the interplanetary magnet field (that comes from the Sun) spreading is the aforementioned heliosphere. Its border is decided by the balance between the Sun's wind dynamic pressure and inters tears' gas pressure, galactic magnet field, and galactic cosmic rays. It is estimated that the heliosphere border is at (50–100) AJ from the Sun, which is far beyond Pluto's orbit. It is estimated that its form is elongated.

And now the real picture of the corona and the happenings in it.

We can see that the antigravitational speed of the Sun's evaporated substance culminates between the heights of 10,000 km and ~70.000 km. In that period, the temperature rises from 104 K to almost 2 million K. Only after that does it start decreasing slowly.

The fact that the light of the emission corona consists of bright lines of the highly ionized atoms of iron, calcium, nickel, and other heavy elements confirms that those are magma va- pour, both light–surface and heavy–subsurface.

In anti-gravitational acceleration, and thus heating, it comes to the breakdown of the molecules (gas) into atoms, but not all the heavy elements break down to the level of α particles and protons. The existence of the areas in the corona, where bursting into flames with intensive emission of X and UV radiation happens, and where the temperature rises to several tens of million K, testifies about the processes of heavy elements atom disintegration to the level of α particles and protons. There, we have fission reactions present.

Active areas in the photosphere cause further effects in the chromosphere, and then they cause them in the corona. That is how flashes, rays, arches, plumes, condensations, and bright spots occur.

Corona's hollows occur above the areas of the photosphere that are not active, most often around the Sun's poles.

We have already learned how and why these things happen. However, protuberances deserve special attention and a detailed explanation.

Protuberances are gigantic eruptions of photosphere volcanos when spirits of magma are ejected high into the Sun's atmosphere. Noted temperature of protuberances actually confirms that they are formed of hot magma ejected from deeper layers of the Sun.

Protuberances' long life, that is, their long hovering, is a simple consequence of buoyancy from the Sun's substance accelerated by antigravitation. The Sun's wind, at its origins, not only compensates for the weight of magma that constitutes the protuberance but also heats it up; that is, it prevents its quick cooling and becoming heavy, and thus, its quickly falling back.

Magma is gravitationally attractive. It keeps itself in a state of protuberance, and it is also attracted by the Sun and thus prevented by the Sun's wind from blowing it into interplanetary space.

With arch-like eruptive protuberances, there is an abrupt rise in the arch size exactly because of the substance pressure (gas) that is anti-gravitationally accelerated upward up to the point when the arch's breakage occurs at its highest point. As magma has to return to the Sun's surface, it happens by its sliding down the parts of the arch.

The anti-gravitational dismissal of the Sun's vapor and gases that begins right at their occurrence, in time and height, turns into the Sun's wind. We have already seen up to which speed and which temperatures.

It is especially important to note the fact that the Sun loses its substance through the Sun's wind. The loss of the substance means the lowering in the gravitational force, which means the reduction of speed and temperatures of the particles in the Sun's atmosphere, which means the extinction of the Sun's glow, and that means doom for us here on Earth. Nevertheless, there's no room for worrying because the loss of the Sun's substance is so scant that the Sun will continue serving us for much longer than the' fusional' anticipated several ten billion years.

Now, let us pay attention to Table 2, which shows us the chemical structure of the Sun's wind and the relative content of atoms in it. Sun's wind's key components are H, whose relative content is

0.96, and 4He, whose relative content is 0.04. It was just these data that led the scientists to conclude that the Sun is a gas ball composed of hydrogen that becomes ^{4}He by fusion. That is how, from correct data, because of the imperfect theory and the ignorance of the true nature of the mass interactions (gravitation and anti-gravitation), wrong conclusions were made.

This chemical structure of the Sun's wind is the consequence of the disintegration of the molecules and heavier elements' atoms that constitute the gases made by cooling magma or ejected in eruptions in anti-gravitational dismissal and acceleration, which is followed by the enormous rise in their temperature.

Cycles Of The Sun's Activity

Observations of the Sun's occurrences have led to the discovery of its cyclic movement. The cyclic movement of the spots' number in time is, on average, 11.2 years. Since the monitoring of the Sun's activity (half of the 18th century) until today, 23 cycles have been noticed. Picture 31 is the results of the monitoring of the Sun's activities from 1749 until today.

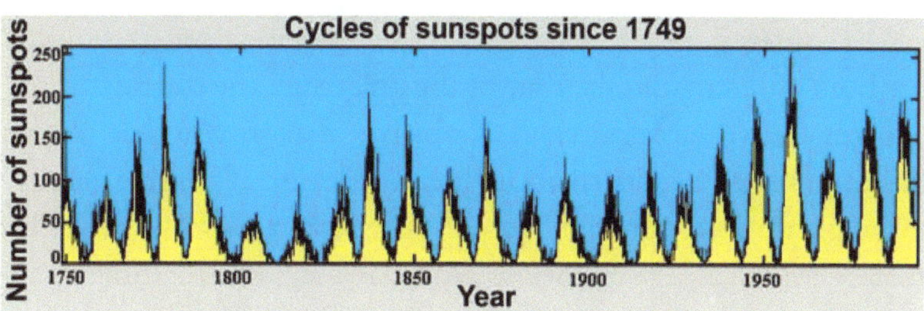

Picture 31.

The minimum and maximum of the spots are clearly noted. The time interval between the two minimums in the spots' occurrence defines the duration of the solar activity cycle. Picture 32 is

109

the change in the number of spots during the 22nd and 23rd cycles of the Sun's activity.

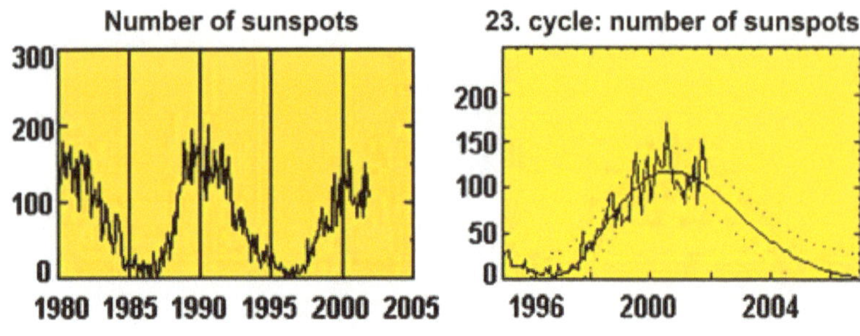

Picture 32.

It is important to mention that not all cycles last equally. For the period from 1755 until 1945, periods between two neighboring minimums varied from 9 to 13.6 years and between the maximums from 7.3 to 17.1 years. The mentioned irregularities occur from cycle to cycle without any notable regularity. Based on the activities in the earlier cycle, all the prognoses about the coming cycle are unreliable.

In contemporary times, based on the Doppler's progressions, it has been noted that in the time of the quiet Sun, the rotation of the equatorial areas accelerates. It is obvious that in the time of the maximum solar activity rotation 'breaks' because of the volcano and magma eruption formation from the deeper layers of the Sun to its surface, which is, again, an indicator that the Sun's rotation is not differential only on the surface by the holographic latitudes, but that the Sun's rotation is differential by the altitude. It is obvious that the 'chemically heavier' magma is rotated slower than the surface layer, which is comprised of 'chemically lighter' magma (picture 33).

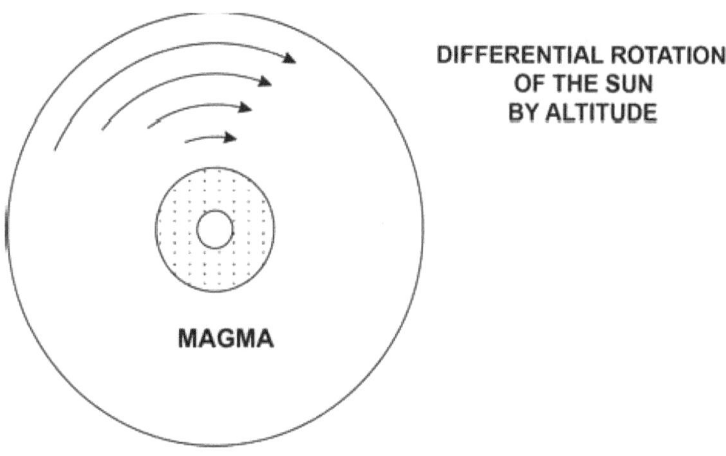

Picture 33.

Apart from the number and the surface of the spots, their distribution along the holographic latitudes during the cycle changes. The first spots in the cycle are formed around 30oN and 30oS, and then they start occurring closer and closer to the equator. In the time of the maximum, spots occur around 15°N and 15°S, and the last spots in a cycle are at about 8°N and 8°S. Rarely can the spots be found at latitudes larger than 45° and lesser than 5 °.

On the 'butterfly diagram' (picture 34), the first spots belonging to this cycle appear at 30o before the last spots from the earlier period disappear at 8o. Such overlap of the cycles lasts about three years on average.

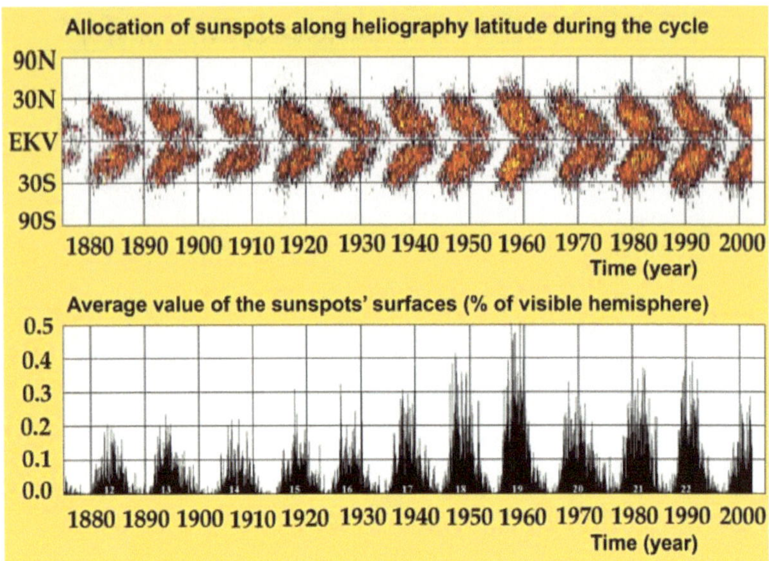

This way of spots' appearance points to the difficulties that are met by the first eruptions of the hotter and 'chemically heavier' magma in the new cycle in breaking out through the quickest layer of the equatorial magma.

As the eruptions get stronger during one cycle, the spurts of magma from the depth manage to burst through the zones of surface magma that are closer to the equator. By breaking through, they slow them down because they function as 'cobs' that stick themselves into one layer after another. As those 'cobs' mostly occur at the time of maximum, it is logical that the deceleration of the equatorial magma is then the highest.

Oscillation of the Sun's activity is obviously manifested not only through the appearance of the spots but also through other forms in the Sun's atmosphere. Growth in the activity is also manifested in the growth of the intensity and the frequency of the chromosphere explosions, with which the flashes, the UV, and radio areas of the electromagnetic radiation are connected. Intensifying the activities also implies corpuscular emission in the form of the solar wind and the Sun's cosmic rays. During the cycle, a change in the division and the number of protuberances occurs: main areas of protuberances' occurrence move to the equator (which is logical, because they follow the spots, that is, they occur in the same areas as the spots), while protuberances of greater heliographic latitudes migrate towards the poles and get there at the time of the maximum of the cycle.

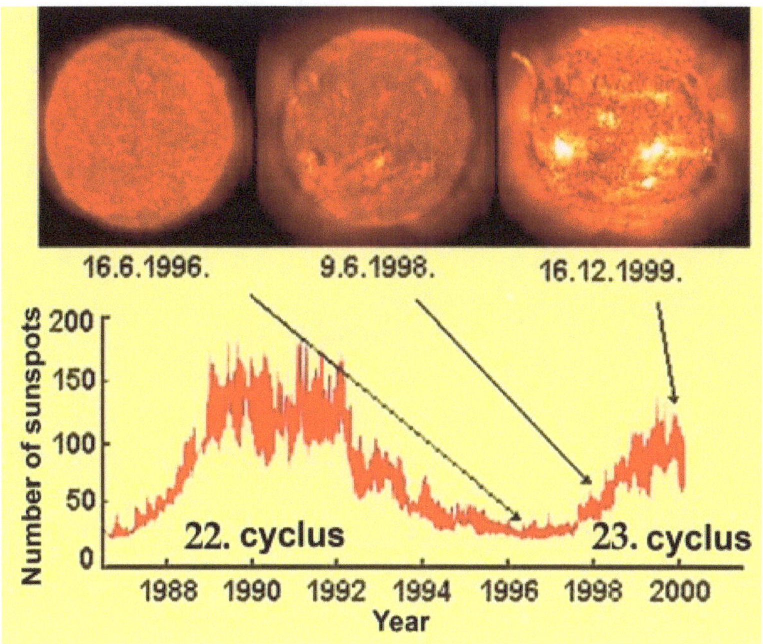

Picture 35.

Picture 35 depicts the development of the 23rd solar activity cycle. Hα pictures show the Sun's' quiet surface' during a time of minimal activity. At the time of the maximum activity (second half of 1999 and 2000), manifestations of the violent activities are clearly visible: we can use multiple explosions, protuberances, and mass ejections.

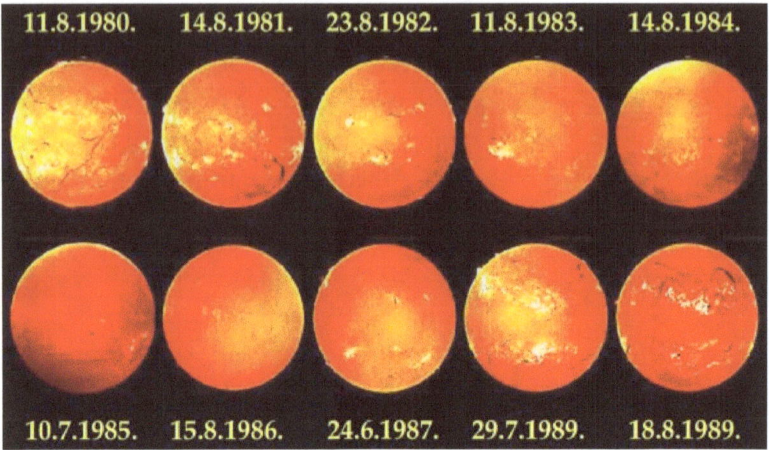

Picture 36 shows the Sun's appearance between the maximums of the 21st and the 22nd cycles.

It is obvious that the temperature changes in the mentioned cycle. It is lowest during the minimum of the Sun's activity, and highest during the maximum of the Sun's activity. What kind of a process could cause such a cyclical temperature change in the whole Sun?

To understand that we must see the Sun's movement in our galaxy.

Our Sun In Our Galaxy (Sun And The Milky Way)

Sun, by its characteristics, stands for an average, dwarf, yellow star. It is thought that 2% of stars belong to this type, which means there are several billion of them.

Galaxies are a gravitationally limited star system. They consist of a large number of stars and the inter-star substance in the form of gas and dust. Depending on the type and the size of a galaxy, the number of stars in them can go from several million up to several billion. Till today several thousands of the brightest stars have been studied. They are the basic structural element for even larger associations in the cosmos – clusters, and superclusters of galaxies.

Our galaxy (the Milky Way) belongs to the class of spiral galaxies, which are distinguishable by their characteristic spiral arms (branches) there are usually two, while the others can be developed at the ends of the spiral system.

Part of our galaxy in the night sky is distinguishable as a bright, pale lane of unequal width that divides the celestial sphere into two parts (picture 37).

Picture 37.

All the stars we see in the night sky belong to our galaxy. A schematic appearance of our galaxy is shown in the picture 38. Observed from both sides it has the shape of two collapsed plates whose diameter is about 30 KPC (100.000 LY).

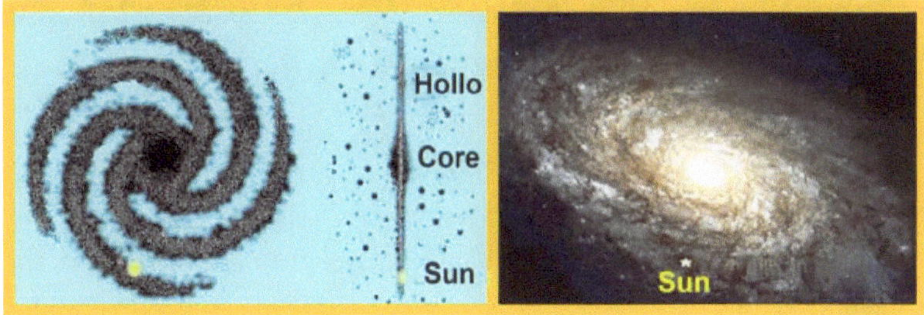

Picture 38.

The thickness of the middle bulge (nucleus) is about 4 KPC (13.000 LY). The middle symmetry level (the galactical level), which lies in the Milky Way and divides the galaxy symmetrically, is easily observable.

Parsec (PC) is an astronomical unit for length, and it represents the distance from which the big half-axis of the Earth's path, which is one astronomical unit, can be seen under the angle of 1" (one angular second). The astronomical unit (AU) is a unit for measuring length in astronomy. It is equal to the big half-axis of the

elliptical path of the Earth around the Sun, that is, by the characteristics of the eclipse, the average distance of the Earth from the Sun.

According to the contemporary measure, 1AU = 149.597.870,5
km, which stands for a length of approximately 149.600.000 km.

The calculations show that it is valid:

$$1PC = 3.262 \, LY = 206.265 \, AU = 30,86 \times 10^{12} \, km$$

Where LY light year is the way, the light passes in a vacuum
within a year ($1LY = 9,46 \times 10^{12}$ km).

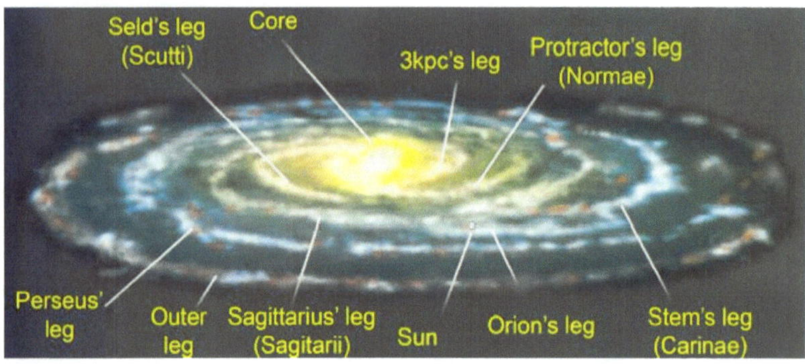

Picture 39.

The Sun is almost at the galactical level from the inside of the
so-called Orion`s leg. The dark space of the night sky is what we see
when we see normally to the galactical level. Based on different
measuring methods and the division of the galactic objects, it was
set up that the Sun is 8 to 10 KPC from the center of the galaxy.
Based on the galactic objects' movement and the analysis of the
radio-radiation that comes from many ways, it is thought that the
most probable distance of the Sun from the galaxy's center is about
8,5 KPC (28000 LY) (picture 40).

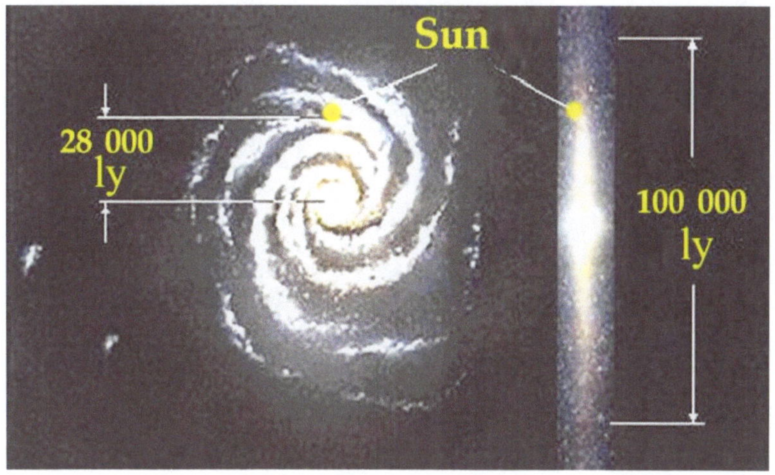

Picture 40.

These acknowledgments finally contradicted the apprehensions that the Sun had a privileged position in the galaxy and in the cosmos. The distance from the Sun to the galaxy's nucleus has favorably affected the development of life on Earth. Namely, the concentration of the stars in the galactic nucleus is huge, so the lethal part of their electromagnetic radiation (UV, gamma, and X) is multiplied intensively by the radiation of the sole Sun or the radiation at the place where it is found.

It is estimated that there are between 100 and 300 billion stars present in the Galaxy. Almost 90% of the visible mass is in the Sun's orbit sphere of the radius, around the center of the galaxy. Objects of the galactic discs rotate on almost circular paths around the galactic center.

The largest concentration of stars in the galactic discus is up to 10 KPC from the center of the galaxy. With alienation from the center, a decrease in this concentration is noticed. We should bear in mind that stars usually appear in doubles, multiples, or clusters.

The disc is encircled by a spheric halo in which stars with a mass of 0.85 M☉ are found. In the chaotically positioned, very elongated elliptic paths, they rotate, mainly around the center of the galaxy, at speeds (50–150 km/s).

The outer part of the galaxy is the corona, which spreads up to 100 KPC from the galactic center.

Our galaxy spins around the axis of symmetry, which is normal to the galactical galaxy clockwise – seen from the north galactic pole.

Analysis of Doppler's movements of the galaxy's spectral lines showed that the objects of the spiral structure (stars, clouds of the inner star gas) move around the center along almost circular paths, but at various angular speeds (picture 41).

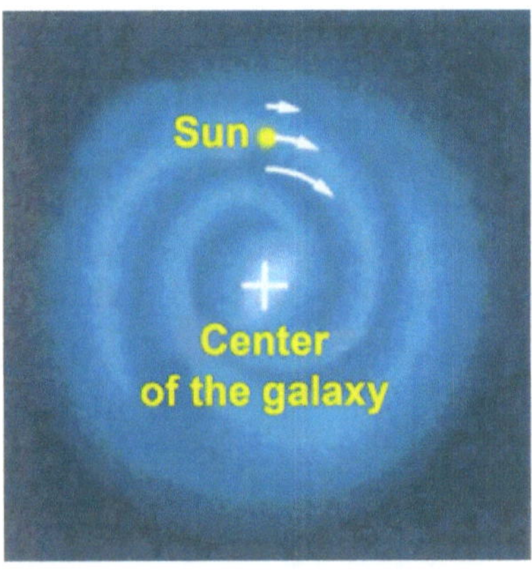

Picture 41.

Angular speed in the central parts of the galaxy is constant; that is, the galaxy rotates as a solid body there. With the alienation from the center, the angular speed of the spiral structure rotation decreases.

The speed at which the Sun is moving around the galactic center was decided based on the movement of the extra-galactic mist that does not take part in the movement around the galaxy. The Sun rotates around the galactic center with a speed of about 230 km/s (828.000 km/h). Even though, from the point of Earthly notions, the world is at an extremely high speed, it takes the Sun 230 million years to make a full circle around the center of the galaxy. This time interval is known as the galactic year (picture 42).

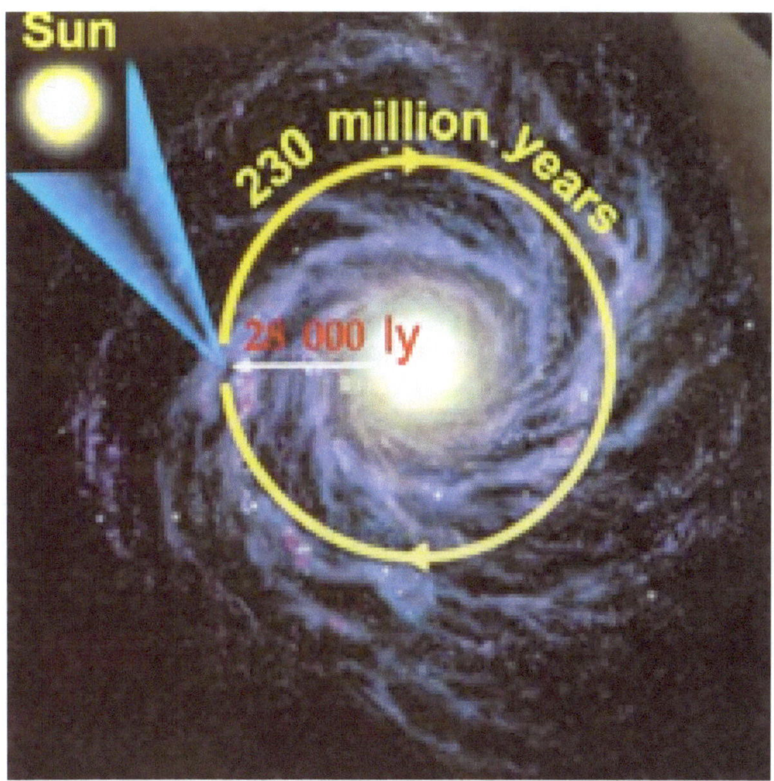

Picture 42.

The Sun is not exactly positioned at the sole galactic level. Today, about the galactic level, it has moved northwest for about 8PC (Oko 26 LY). This discrepancy on the galactic distance scale is negligibly small. However, because of that position, during its circular movement around the Galactic center, it seems to swing up and down (picture 43).

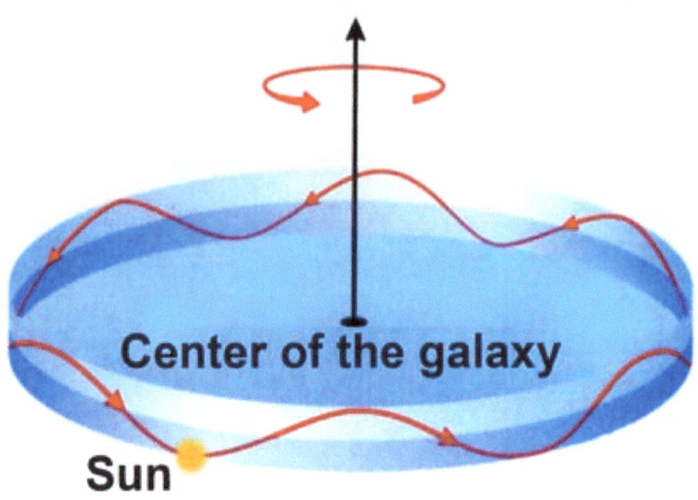

Picture 43.

Research shows that the Sun periodically passes through the galactic level. Similar movements are typical for other stars that are in the vicinity of the galactic level. The cycle of this kind of the Sun's swinging in orbit around the galactic center is about 33 million years.

It seems logical to me that this kind of movement is the consequence of the Sun's rotation around the lengthwise axis of Orion's leg, where the Sun is located (picture 44).

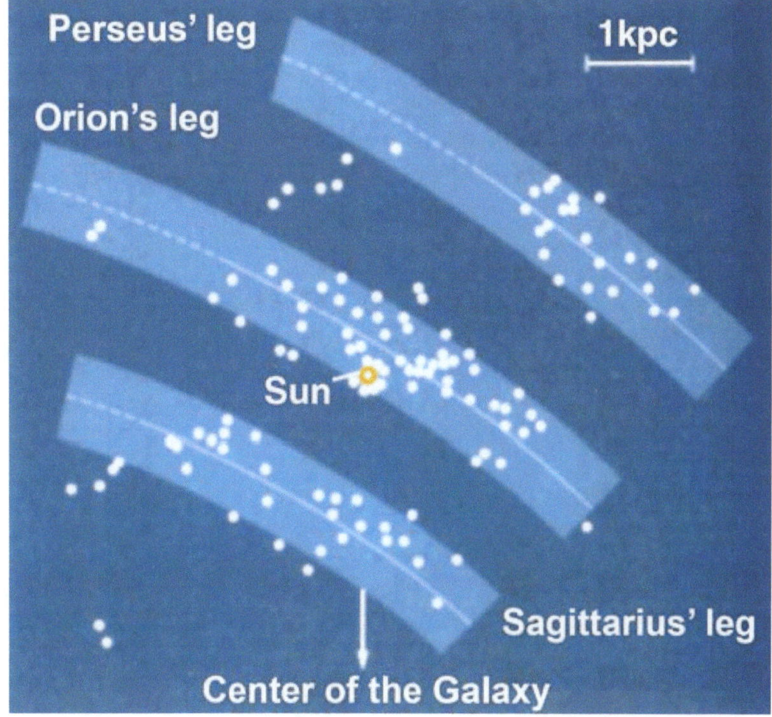

Picture 44.

It seems logical to me that this kind of movement is the consequence of the Sun's rotation around the lengthwise axis of Orion's leg, where the Sun is found.

That would mean that the Sun moves around the center of the galaxy along a spiral or a closed helicoid.

within a circle. From the data provided it is noticeable that the Sun rotates around the lengthwise axis of the Orion's leg 3.5 times, that it passes through the galactic level 7 times until it makes a full turn around the galactic center. While turning around the lengthwise axis of Orion's arm, the Sun partially approaches the galactic nucleus, and on the other, it moves off from the galactic nucleus, going through its closest and farthest position about the galactic nucleus. Since that movement of the Sun is happening along with the unavoidable galactic wind and radiation, at the same time, it means that the Sun would change its temperature depending on its position and its movement. When it is closest to the galactic nucleus and when it is passing through one galactic level, its temperature will be much higher than when it is on the opposite side, when it is farthest from the galactic nucleus.

When it is between the two positions, its temperature will vary as well; when it is moving away from the galactic nucleus and when it is 'running away' from the galactic wind its temperature will fall, and when on the opposite side it starts approaching the galactic nucleus and 'running into' the galactic wind, its temperature will rise.

That automatically means that by that its activity will change. But those are terribly slow changes – as we have seen one cycle lasts about 66 million years.

But what could be the reason for the 11.2-year cycle?

The first reason could be that the intensity of the galactical radiation and the galactic wind is changing by the same cycle.

The second reason could be that the Sagittarius arm baffles or shields the Sun in such a cycle (look at picture 44). Those would be, let's say, partial eclipses of the galactic nucleus. If our Orion arm rotates around its lengthwise axis, then all the other arms do, too, as well as the Sagittarius arm, which is closer to the galactic nucleus and can shield the galactic nucleus in different manners.

The third reason could be the Sun's movement along the spiral that goes around the big spiral, which would cause the existence of another star that makes a binary system with the Sun.

Further close observations will surely lead us, one day, to the discovery of what exactly is the cause of the noticed cycles of the Sun's activities.

However, I would like to emphasize here that the movement of the stars significantly influences their life and, as we are about to see, their destiny.

SOLAR SYSTEM CREATION

When the star is so small that can't create another generation of stars, it creates its own stellar system. It creates planets. Our Sun is that kind of small star and it creates planets. We can see the history of the life of the Sun by seeing the order and size of planets. The first 4 planets are Rocky planets. Then we have an asteroid belt between Mars and Jupiter. And then we have Jupiter, Saturn, Uranus and Neptune. We now consider them like gas giants. But they're not. They're big spheres of liquid lava. So, our Sun was trying to make another star, but it couldn't; Jupiter was the best shot or the best try for our Sun to make another star. Jupiter has a liquid surface. That's the reason why Jupiter has such a dense atmosphere with the clouds and all kinds of things we see. The other big planets in our Solar System are the same. They have liquid lava bodies. Otherwise, they will not be able to hold their satellites around. The logic is back for them.

So, from eruptions from the surface of the Sun and especially from the equatorial plane, the planets came into existence. The asteroid belt came into existence, and the Kuiper belt also came into existence from eruptions from the surface of the Sun. We must remember always that solar wind blows away all the products of the eruptions from the Sun's surface. But from the eruptions from the

North and South Pole of the star, either our Sun or the other small star, due to pulsations, many of the asteroids are created and they end up in the Oort cloud. They are rejected much faster, and they will reach a wider orbit than any other body created from the star, so they will create an Oort cloud. We now understand how many rocks are around every small star, which cannot create new stars, so they have planets, a Kuiper belt, and an Oort cloud. The logic is back; around boiling lava, stars are Rocky planets, asteroids in belts, and clouds. A remarkably interesting thing is that planets, during their formation, create hollow inside, as stars do, but they end up with the openings on their North and South Pole. So, all stars are hollow, but planets are hollow as well.

Globular and Spiral galaxy Creation

Now, if we consider big stars, which are older generations than small stars, which still create smaller stars, we can easily absorb the logic that was given in the previous chapter. And we can understand how Globular Galaxies are created. The big Mother stars, in the center of the Globular Galaxy, create their system in a similar way to small stars, but instead of the Kuiper belt and Oort cloud full of meteorites, they will have their Kuiper belt and Oort cloud full of stars, and it will make their Globular shape look. So, we have a pulsation of the mother star of the Galaxy, which means new stars will come into existence. At the surface of their globe will be the younger stars and inside their globe will be older stars. That is the pattern of the Globular Galaxy, and the James Web telescope has seen that. We have that now like a fact. Now let us consider even bigger mother stars of the Galaxy, which can create Spiral Galaxies. They are much bigger. Their boiling lava body is much thicker, and pulsations are rare. The frequency of the pulsations is low, but during their pulsations, a huge amount of matter is ejected in their jets, and those jets create the legs of a Spiral galaxy, which is full of stars. The best example is our Milky Way galaxy. Going further up in size of the star, we can now understand how more complex structures are created, like clusters and superclusters of galaxies.

HUBBLE AND JAMES WEB TELESCOPE

What the installation of Hubble and the James Web telescope has presented to us is totally in contradiction with everything we assumed that we know about the Universe. The James Web telescope discovered galaxies that are so old that they contradict The Big Bang Theory. Also, it has been discovered that the Universe is not isotropic, and it is not expanding in all directions the same way.

We also must consider the Red Shift or Hubble constant. Because the universe is pulsating like everything else, it's now in the process of ever-faster expansion because of ever-increasing temperature or energy coming in.

We must consider that when we look further away, we look at the colder Universe and its reason for the Red Shift to be reconsidered and the temperature of the stars, galaxies, and every other structure we see due to temperature at that time. So, we must reconsider everything we thought we knew. The main thing for us is to understand that we know nothing. If we can understand that then we can grow in our knowledge. Step by step using the help of Heart Science.

A Star's Movement Influences Its Life and Destiny

To correctly understand the influence of the movement of a star on its life, I will remind you of the phenomena of the star wind.

As we have already seen, star wind is a consequence of the star's anti-gravitational rebuffing of the evaporated photo-spheric substance. The temperature of those molecules rises with acceleration – that is what we already know from the molecule-kinetic gas theory: higher speed implies higher temperature. However, the molecule-kinetic theory of gas has not explained the true reason for that. How can temperature cause speed? How can speed cause temperature?

The necessity is the existence of the gravitational field. Everything that the molecule-kinetic theory has described happens in the gravitational field of the Earth. A star's wind is born in the gravitational field of a star.

Body temperature is the factor that changes – we have already seen how – the quantity and the quality of the body's mass.

Change in the speed of a body implies the existence of acceleration. Acceleration implies the action of power.

In the gravitational field of a body, gravitational force attracts all the bodies with attractive mass, and the anti-gravitational force rebuffs all the bodies with negative mass.

Therefore, the temperature of a body, which is in the gravitational field of another body, decides whether that body will be gravitationally attracted or anti-gravitationally rebuffed.

Since the molecules of the evaporated photosphere substance have negative mass, they are rebuffed from the star by the anti-gravitational force, and thus, their speech rises.

With the rise in the speed, their temperature rises, and thus, their mass negativity quantitatively rises, and an even larger anti-gravitational force of rebuffing affects them. That is how it comes to larger and larger speeds and higher and higher temperatures of the star wind with the alienation from the star's surface.

The reversed dependence of the anti-gravitational force of the square distance, of course, leads, at one point, to reach the maximal speed and the maximal temperature, after which a gradual fall of both the temperature and the speed follows.

Hence, temperature, through the anti-gravitational power, enhances the speed at which the molecules of the photosphere vapor alienate themselves from the star. That is the answer to the question of how the temperature enhances the speed.

Now, we have to completely dismantle the mechanism of how the speed enhances the temperature.

In fact, the speed is not the factor that enhances the temperature; the factor is a change in the speed or acceleration that is the consequence of the action of the anti-gravitational power. What is really happening during the action of a power to a body?

This is an utterly fundamental question in physics, and the answer to it must be completely understandable and logical.

When a power affects a body, it exerts activity upon it. The activity that a force exerts upon a body is divided into three parts. The second part is used to change (enhance) the kinetical energy of the body because its speed has changed (enhanced). The third part is used to change the potential energy because its position in the power field is changed. The first and most exciting part for us is used to overpower the body's inertia.

What is inertia? Physics says that it is a body's feature to resist the change in the state of its movement. It is logical that that feature of a body lies in a certain physical reason. If a force affects a body, and a body resists the effect of that force, it logically means that there is a force of an inverse action.

Since I am going to work on this in detail later, now I will just say that inertia is a consequence of the interaction of the physical body with the physical space. To make myself completely clear, I will say that there is friction between the physical body and the physical space.

Where there is friction, there is heating, that is, the change in temperature.

Therefore, the first part of the action that a power is exerting on a body is spent on overpowering the inertia, that is, the friction between the body and the space, which causes the change of the inner energy of both the body and the space. Here, we are interested in the change of the inner energy of the body, which means, in the outcome, that the temperature of a body rises on account of that part of the action of the power over the body that overpowers the energy.

The law of conservation of energy is now completely satisfied.

So, antigravitational power exerting action over the molecules of the evaporated photosphere substance uses one part of that action to raise the temperature of those molecules due to the friction between the molecules and space. And that is the essence of the answer: how a change in speed causes a change in temperature.

Let us remind ourselves of something else. When the speed of a body is unchangeable, both for its direction and its intensity, we say that that's inertial movement. For inertial movement, the rule is that the speed is constant, which means that the temperature is constant, too.

But... nowhere in space do we have internal movement.

Everything is spinning around itself and something else.

The non-inertial movement implies rotation because the direction of speed is being changed there, although the intensity stays the same. Rotation is a consequence of the existence of the centripetal force, and gravitation plays its role in the universe.

Therefore, all the bodies that spin around themselves or around another body are in permanent friction with the physical space and in a permanent process of heating.

All that we have just said applies to stars, as celestial bodies, too.

The spinning of a star around its axis, as well as the spinning of a star around the mass center of a double or multiple system, and the spinning around the center of a galaxy, causes heating of the star.

The existing astrophysics does not know this!

New astrophysics must include this mechanism of rise of the stars' temperature.

If we name the factors that decide the temperature of a star, then it looks like this:

<u>Gravitational shrinkage of a star</u>. It is decided by the amount of the substance and its temperature. That overall mass shrinks the star gravitationally until there is a balance with the force of the anti-gravitational rebuffing, whose outcome is in the central hollow of the very star. The larger the attractive mass of a star is, the larger the shrinkage is, along with the pressure density, and temperature of a star. Then, that means that the 'vaporization' of a star through thestar wind is stronger; that is, the star loses its substance quickly. When a considerable substance loss leads to the weakening of the gravitational shrinkage, the star's wind will die away, and overall, the star will lose its glow (luminosity). Hence, the star's higher temperature means a stronger glow (luminosity) but a shorter life span. Of course, the larger the starting amount of the star's substance, the longer its life and its glow.

<u>Star's radius and the angular speed of the spinning around its own axis</u>. The more significant the amount of the substance a star has, the longer its radius. A larger radius causes greater gravitational shrinkage – higher temperature, and so on. The same angular rotation speed around its own axis for two stars of different radiuses will cause larger heating for the star with a longer radius. But that heat-

ing, due to the spinning around its axis, can provide greater glow (luminosity) to a star if it spins faster than another star that spins slowly, while they are of the same size that is the same substance amount. Along with that, the star's life span will be different, too. The different rotation of the sole stars is also a consequence of the friction of the star and the space.

Distance from the mass center system and the speed of spinning around it. Stars usually appear in binary or multiple systems. That means that a star that spins faster around the mass center of its system will have higher tee premature than the same star that spins slower around the mass center of a similar system. Lesser distance from the mass center means quicker rotation of the entire system and larger heating of the star under the influence of the star wind of the other system members.

Distance from the galaxy center. We have seen that the galactical nucleus rotates as a solid body and at a larger speed than its arms, which leads to the spiral look of galaxies. The farther the star from the center of the galaxy is, the lower its rotation speed around it is, and its heating is therefore lower. Con- The concentration of the stars rises with approach to the galactic center, which means that the galactic wind is getting stronger, that it heats the stars that are closer more than those that are farther away.

Hence, a star closer to the galactic center will have a higher temperature and glow (luminosity) than the same star found farther from the galactic center. The intensive galactic wind in its nucleus, which originates from the star winds of the galactic nucleus, is a factor that, with its anti-gravitational nature, expands the galaxy and separates the stars between themselves.

Galaxy movement in a cluster and a supercluster of galaxies. A galaxy spins around the galactic axis, but it also moves at great speed around the center of a cluster of galaxies and a cluster of galaxies around the center of a super-cluster of galaxies. Superposition of those movements with the star movement (around its axis, around the system center and the galaxy center) heats the star additionally because of the great speeds at which the sole galaxies move.

<u>Star's chemical structure</u>. Since not all stars have the same chemical structure, that means that some of them are made of heavier and some of lighter substances. The heavier substance causes greater gravitational shrinkage, and, during their movement, it conducts greater friction with space, and it logically implies that heavier stars have a higher temperature, higher luminosity, and more intensive wind than the 'lighter' stars of the same size and the same conditions of movement.

Also, the substance of the 'heavier' stars is disintegrated in a different manner in the star wind than the substance of the 'lighter' stars, so according to the chemical structure of a star's wind, we can conclude about the 'weight' of that star, that is, about its chemical structure.

<u>Intensity of the star's wind</u>. When all the other factors do theirs and get a star to a certain temperature, the rising wind continues to additionally heat the star because a part of the heat of the star's atmosphere falls to its photosphere and heats it up even more.

So, we see that numerous factors affect the maintenance of a star's temperature, thus affecting the length of its life. Such mechanisms provide stars with a much, much longer life than expected by existing astrophysics.

Origin Of Stars

The question of the origin of stars is a fundamental question in astrophysics. To find the answer, I will begin this investigation by repeating a paragraph I had previously written. It is a paragraph on galaxies, and it goes like this:

"Galaxies are gravitationally limited star systems. They consist of many stars and interstellar substances in the form of gas and dust. Depending on the type and size of a galaxy, the number of stars in it can go from several million to several thousand billion. Until today, several thousand of the brightest galaxies have been studied. They are the essential structural element for even larger associations: the space clusters and superclusters of galaxies. "

Hence, the place of the origin of stars is a galaxy. But a galaxy is a pretty big space. Where exactly do the stars arise?

Obviously, it is there where the concentration of the stars is the largest, and that is the galactic nucleus.

We have already seen that the galactic nucleus rotates faster than the outer part of a galaxy and that it acts.

compactly, as a solid body. Gravity and anti-gravity are the forces that make it so compact. Gravity prevents the dissipation of stars, and anti-gravitation prevents the collapse of the nucleus. We can see that anti-gravity is even more dominant because all stars move off from each other despite the gravitational at Traction. Anti-gravitation, which is the source of the star winds, therefore the galactic wind, extends not only the galaxy but the whole universe, too.

That means that anti-gravity dominates both in a galaxy and in a galaxy and in a galaxy and in a galactic nucleus. Hence, in the heart of the galactic nucleus, there is a huge star. That star can be named 'galactic mother'. Galactic mother gives birth to the whole galaxy – it is the mother of all the stars in that galaxy. Of course, if a galactic mother creates too big stars, then those big stars create smaller stars. That is the mechanism of origination of binary, multiple star systems, and star clusters.

So, stars originate from more prominent stars; more giant stars create smaller stars. That is the way!

A question arises—how far can it go in both directions? In the direction of star shrinkage, it goes until the stars are no longer able to create new stars. When that happens, the stars create planets, comets, asteroids, and all other things that make a system around a star. I will write about that later.

In the direction of star growth, there also must be a limit. If the galaxy stars were born by the galactic mother, then logic says that the numerous galactic mothers were created by the mother of the galaxy cluster. Numerous mothers of the galaxy cluster were created by the mother of the galaxy supercluster. The Cosmic Mother created the numerous mothers of the galaxy supercluster.

Cosmic Mother is the first-formed star that is used to combine the whole matter of the cosmos. That considerable ball (of the whole) matter was heated till the incandescence by the gravitational shrinkage, and at one point, it started the creation of the cosmos through the process of creating stars smaller than itself.

A different scenario is possible. Not one, but several gigantic stars (MSSG) might have originated from the whole cosmic matter, which, after their incandescence, started creating the cosmos as we see it today.

A notable example of the origin of stars is star clusters. They form a large number of stars, varying from several thousand in the case of open clusters to several million in the case of globular clusters. All those stars were born at approximately the same time; they are of the same age and have the same chemical structure, which is metallicity.

A big star that is the mother of a star cluster was heated to the point of boiling. When it boiled – and boiling is vaporing of the value – due to the anti-gravity, an explosion.

of the surface layer occurred.

From the scattered magma, gravity formed new, smaller stars that went on living in the gravitationally limited system that we name a star cluster. Of course, those stars' winds cause the star cluster's expansion by the general expansion of the cosmos.

"Dating of the globular clusters is especially important in astronomy because they are the oldest objects in the universe. Their age is today estimated to be 13 to 16 billion years, which confuses contemporary cosmology: that number is larger than the generally accepted estimates of the universe's age. The solution to this problem is in sight today, but it could seriously shatter the theory of star evolution and the cosmological models. "

This quotation is from the book, „Birth, life, and Death of Stars, "whose authors are Nicolas Prancos and Tierry Mont-Melle.

It confirms what I am saying—that the universe is much older than we think because stars live longer than we think. The wrong idea of fusion as the source of the stars' energy led us to the wrong determinants of the stars' duration and, therefore, to the wrong estimate of the whole universe's age.

Now, I would like to work in more detail on the question of the stars' chemical structure, which is their metallicity. We can see that there are stars of different chemical structures; they are formed of magma of different' weights.' What and why is happening there?

To understand that we are going to analyze a big star that gives birth to a generation of smaller stars. Whatever its chemical structure is, there is a universal validity. Since the star is incandescence but liquid boiling magma, the layering of Magma by chemical weight occurs in it. The surface layer has the lightest magma, and as we go depth ward, layers are made of heavier and heavier magma.

The last layer, or the layer around the central hollow in the star, is made of the chemically heaviest magma. All layers are under pressure, the inner ones are under the pressure of the layers above them, and the surface layer is under the pressure of the inner layers.

The chemical weight of each layer's magma, combined with the pressure under that layer, finds its point of boiling. Of course, the surface magma layer, which is the lightest and under the smallest amount of pressure, has the lowest point of boiling.

When the surface magma layer boils, it is ejected in the anti-gravitational explosion. Stars born from that layer's scattered magma will be in the class of the lightest offspring. The explosion of the surface layer causes tightening, that is, a rise in the pressure and, therefore, in the temperature of the layers beneath the surface.

Until now, around the bare star there is a whole class of the lightest stars, and there is a clash of the star winds of the mother star and the daughter stars. That is what causes further alienation of the daughters, but the existence of the pressure on the mother star, as well. As the daughters move away from the mother, the pressure on the mother starts the new surface layer will weaken. At one-point, new conditions for the existing mother star's surface magma layer will occur, and that will cause its ejection in an anti-gravitational explosion. Stars that are created from the scattered magma of this layer will be in the class of a bit heavier star offspring.

The process is repeated after the explosion of the first layer, and when conditions for the new boiling of the newest magma layer are reached, it will explode, and then a new class of even heavier star offspring will be created.

In this manner, after various periods of time, layer after layer will explode, creating heavier and heavier classes of offspring stars until the rest of the mother star is not 'thin' enough that the boiling of the surface magma layer cannot be achieved. That is how, after a series of substance ejections, the mother star will reach its stability and enter a relatively quiet period of its life.

Such a mechanism of origination of star offspring classes different in weight in approximately concentric spheres should be detectable by observing star clusters.

Of course, in all those explosions of the layers of star magma, we should not expect mathematical precision and symmetry. Physics is a science that describes reality that is around us, and there are always discrepancies with the ideal expectations and predictions.

A realistically possible scenario is the following: because of the differential rotation of the sole surface magma layer on the star, boiling of magma occurs, firstly in the equatorial zone, which would lead to explosions only in that zone and the spreading of the star offspring in the equatorial level of the mother star. Only after that would the boiling and exploding of the complete surface layer occur, and that would severely symmetrically eject the star offspring.

Such a form of spreading the offspring stars is just what we have with galaxies that are flat. They expand at the galactic level, which is the equatorial level of the galactic mother.

This logic leads us to the conclusion that the stars from the galaxy's border should be made of the lightest substance and that, as they approach the galactic center, the weight of the substance rises.

The same should apply to star clusters, while with clusters, we have symmetrical and not level expansion.

CAUSE OF ROTATION OF CELESTIAL BODIES

For us people, it was not difficult to note that the Sun and the Moon rotate around the Earth. And then we established the same for the stars. And then, the picture was spoiled by Copernicus with his book "New Astronomy," in which he explained that the Earth spins both around its axis and the Sun. Only the Moon rotates around the Earth.

All other astronomical observations have shown that all celestial bodies spin around their axis and from a certain center of rotation. Simply upward, rotation is a universal law in the universe. And that law must rely on a certain cause.

If I were kidding, I could say, 'Rotation of the celestial bodies is a consequence of the feature of those who quench their curiosity by constantly wandering and looking at what is going on around them.' Good one, isn't it?

We have already seen that stars arise from larger stars. The rotation of the offspring stars is much easier to explain if the ancestor star is already rotating itself. But how did it come about that the stars rotated in the first place?

Did the Cosmic Mother spin around its axis? Or did the supercluster mothers spin around their own axes?

I can say the following: the inhomogeneous distribution of the matter in the space around the stars, during the formation from the cosmic matter available that they used to attract with powerful gravitation led to that kind of falling of the matter on them that led to the spinning moment. Therefore, we should suppose that the matter's falling was in such a manner that it favored the spinning moment in one direction. However explanations like this are not acceptable. I will begin with the assumption that the formation of either the cosmic Mother or the supercluster mother was such that it did not lead to spinning around its axis.

Therefore, this is the situation. A large amount of the piled matter is heated due to gravitational shrinkage and leads to the forming of the layers of incandescent and liquid magma whose weight is raising going from the surface toward the center. A star that arises in that way has no movement, neither around its axis nor anything else. When gravitational shrinkage leads the surface magma layer to the point of boiling, the first anti-gravitational explosion will occur with the ejection of the first layer's magma into the surrounding space. All those parts of magma will be ejected in radial directions spherically symmetrically, that is in all directions in space.

All those parts of magma had been given linear accel- elation (linear speed) by the explosion.

It is logical to assume that before forming the spherical shapes of the parts, which occurs due to the gravitation action, their shape was arbitrarily irregular. The period in which a piece of ejected magma is in an irregular shape is of the es- essential importance for our analysis. Therefore, we have irregular form magma that is alienating from the ancestor star and has linear speed. What is happening then there?

Multiple actions are taking place.

The first thing is that the ancestor star's wind exerts different pressure on various parts of the irregular magma. This causes the formation of the coupling forces that start spinning around the mass center.

The second thing is that the star wind and the ancestor star radiation heat the parts of the irregular magma unevenly. The sunny parts heat more than those in the shade, and the thicker parts heat more than the thin ones, because of the difference in the size of the receiving surface.

The third thing is that the thicker and thinner parts of the magma cool down differently – thicker and slower than the thinner ones. The fall in the temperature causes a rise in the mass. A rise in the mass means the rise of inertia, and the inertia rise implies larger friction with the space. The various friction with the space of the various parts of the irregular magma forms the cooling of forces that start spinning around the axis through the mass center. Because of the different masses of the different irregular magma parts, acceleration, that is, the consequence of the explosion, provides different speeds to different parts (to the smaller parts at greater speeds than to the larger parts). That also creates a coupling of forces for spinning around the mass center.

All other actions are superposed, and the irregular-shaped magma starts spinning around its axis and goes through its mass center.

Of course, this situation lasts only for a certain period because gravitation acts, and the irregular magma shape is turned into a ball. Now we have the effect that is known to us from ice-skating. An ice-skater starts spinning around him or herself with his or her arms and one leg spread (the other leg is the axis of the spinning). There the angular speed is not large. But, when he or she puts the arms next to the body, a rise in the angular speed arises. That is the consequence of the maintenance law of the impulse moment.

The same thing happens on the occasion when magma changes from irregular into the shape of a ball. That transformation is causing the rise in the angular speed.

That is how, from an ancestor star that was without linear movement and spinning around its axis, we got offspring stars with linear movement and spinning around its axis.

The newly created situation looks like this. The ancestor star that was never moving and it did not rotate around its axis is now spherically symmetrically encircled with its offspring stars that move radially linearly in their alienation from the ancestor star, and at the same time, they rotate around their axes. At that, the smaller offspring stars will have greater both the linear and the angular speed of rotating around their axes from the larger offspring stars. Let us notice that the offspring stars do not rotate around their ancestor star. When will that occur and how?

Rotation of the offspring stars around their ancestor star will occur when the stars with rotation around their axes start creating their offspring.

What does it look like?

Let us observe one star that moves linearly and spins around its axis. Its linear movement was at the beginning accelerated, and then it turned into a movement with constant speed. That means that linear movement was only at the beginning, a factor of its added heating. After the rotational movement around its axis begins, it becomes a constant factor of the added heating of the star because that is a non-inertial movement (direction and way of the speed continually change).

During its rotation around its axis, a star has the highest peripheral speed on each equator and the lowest on its poles. That means that the surface layer will be mostly heated in the equatorial zone and that magma boiling will occur easiest and quickest there. Explosion of the equatorial zone leads to the ejection of the magma parts in the star's equatorial zone.

Therefore, a star that spins around its axis will create its offspring in its equatorial zone. Expanding of the offspring stars will occur in that zone. But if during the ejection of the equatorial zone magma, each part of that magma has an already existing peripheral speed that is normal to the direction of the alienation speed from the ancestor star. The superposition of the two speeds leads to the star's movement along a spiral path in the equatorial zone around the ancestor star. Distance between the arms of the spiral shrinks in time and turns into an ellipse along which an offspring rotates around an offspring. Of course, this offspring will spin around its axis, too. And so, from an ancestor star that had motion in a straight line and rotation around its axis, we came to offspring stars that spin around the ancestor star along elliptic paths in

its equatorial zone and spins around its own axes.

The elliptical paths of the offspring stars, while rotating around the ancestor star, are the consequence of the ancestor star's movement, either linear or elliptic (around its ancestor). An ideal circle, as a curve, along which an offspring rotates around its ancestor, would be possible only in the case when the ancestor does not have any other movement except the spinning around its axis. Of course, the higher the ancestor's speed, the flatter the ellipse along which the offspring rotates will be, that is, the difference between the ellipse's half-axis will be longer.

I am emphasizing once more that a star that does not rotate around its axis does not create its offspring spherically symmetrically around it, and the star that spins around its axis creates its offspring evenly symmetrically in its equatorial level and that it rotates around it. If we apply this logic to what we see in the universe, we can conclude that galactic mothers are stars that rotate around their own axes. We can also conclude that globular clusters, as the oldest and the farthest known objects in the cosmos, came into being from stars that did not rotate around their own axes. And those stars that did not rotate around their own axes are the first- formed stars, and they started forming the cosmos as we can see today.

Celestial Bodies Rotation Preservation

We have seen how it came to the complete rotation of the celestial bodies, both around their axes and around their ancestors. The question that logically imposes is how those rotations are preserved.

Rotation is, by its nature, a non-inertial movement and it is the source of the added heating because of the friction with the space. That continual friction with space should, in time, lead to the stopping of the rotation, but we can see that that has not happened. Despite the great age of all celestial bodies, they still spin both around their axes and around their- ancestors. What is that mechanism and how does it work when it provides the preservation of the once started rotation?

At the basis of everything lies the process of the universe heating. After the first star's ignition, the process of heating started. With the multiplication of the stars' number, the heating process intensified, and with it, a process of the universe's expansion began. All bodies expand in heat, and so does the universe.

Heating, through the anti-gravitation action, alienates the first-generation offspring from the ancestor, but it also alienates them from each other. Then, new generations of offspring are created, and the pattern repeats itself – alienation from both the ancestors and from each other.

During the first generation of offspring, rotation around their axes occurred, and with the second offspring generation, spinning around the ancestor occurred.

The ancestor that is moving linearly along a curve drags its offspring (that rotate around it) with itself, accelerates them, and turns their paths into ellipses. That is how the very ancestor movement preserves or even accelerates the rotation of its offspring around itself.

When we see the offspring that spins around the ancestor, to whose radiation and wind it is exposed, we can logically conclude that its radiated side is a bit warmer than the one that was in the shade. A difference in the temperature means a difference in the mass. The cooler side is heavier, that is, more inert, and it has greater friction with the space than the warmer side. That inequality of friction with the space of the cooler and the warmer sides of the offspring always creates an added spinning moment that preserves the offspring's rotation around their axes. Continual added spinning movement creates an added gravitational action of the ancestor to the offspring's sides in a way that it is more drawn to the offspring's side that moves from the shade to the radiated part, from the side where the radiated side moves into the shade. The cause is the difference in the sides' temperature and, along with it, the mass difference (see picture 45).

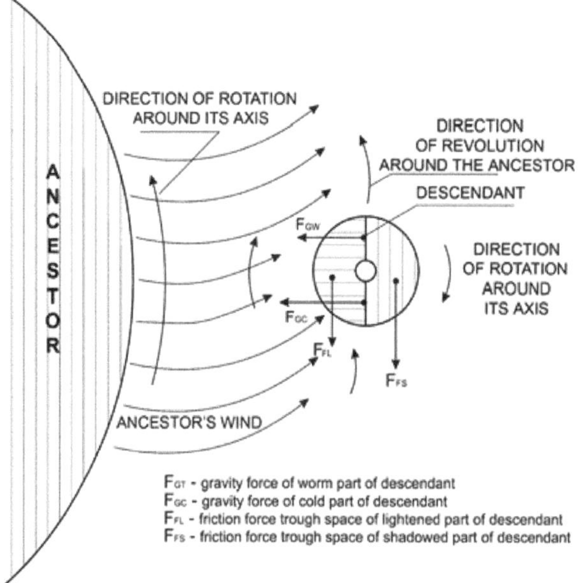

Picture 45.

In the picture we see that the ancestor's 'wind-blowing' effect is such that it contributes to the offspring's rotation around the ancestor preservation and the offspring's rotation around its axis preservation.

The logic of the events tells us that the privileged direction of the ancestor's rotation around the off springs is identical to the direction of the ancestor's rotation around its axis. The privileged offspring's rotation around its axis direction is the one that is opposite to the ancestor's rotation around its axis. Whichever the case, from the knowledge about the rotation directions of the ancestor around its axis and the offspring around its ancestor and around their respective axes, we can draw conclusions about the ancestor's movement through space.

The very process of the universe's heating leads to the heating of all the celestial bodies and, because of that, to the fall of their attractive mass. The law of impulse and impulse-momentum preservation provides that the mass reduction is compensated with an adequate rise in speed, both linear and angular.

That is how the stated mechanisms provide the celestial bodies' rotation preservation from their arousal till today, and they will continue preserving it in the future.

Dimensions and "Constants"

Existing physics has complicated our lives a lot by introducing new dimensions. Three dimensions were not enough, nor four, and it all went up to the number of dimensions a specific theory needed. That is how it goes in mathematics. Mathematics deals with n-dimensional spaces. In mathematics, we can create whatever virtual world we like.

Physics aspires to explain the world around us, the real world. At least, that is how it was at the beginning, and I certainly hope that is how it is now.

On the topic of space dimensions, I would like to say that we need three dimensions for the quantitative space description. Only three. Not more than three. Therefore, three. For the qualitative description of the space, we can in-produce as many dimensions that we need or as many as we can differentiate.

Time is a qualitative dimension; it shows us the quality of the space around us or of the space we want to explore. The introduction of time as the fourth quantitative dimension was wrong; that was an expression of not understanding the quantity and qualities of space.

All 'constants' we have defined in physics are qualitative dimensions of space. They tell us about the quality of the space we are in.

The most important thing I would like to say on 'constants' is that they are not constant. What we have been considering as 'constants' were not constants; they are changeable sizes. Indeed, their changes are terribly slow and exceedingly small.

The second important thing I would like to say about the 'constants' is that they are not universal. Values that we get by measurements are the indicator of a part of the space we are in and where the measuring took place. In some other parts of space, their values would be different.

Let me just explain that with specific examples.

Let us see the gravitational 'constant' γ and the speed of light c. At this stage of universe development, when it is heating and expanding faster and faster, the γ is decreasing, which means that the intensity of the gravitational interaction is weakening.

At the same time, because of the heating, the dielectric and magnetic vacuum permeability value means that the speed of light is increasing. If you think that our lives will become more complicated because of these statements, I will tell you that our life in physics is as complex as it is, and this can only lead to its simplification.

But I admit there will be a turning over in physics, astrophysics, and cosmology. So far, we have been deciding and checking the 'constants' for the most correct value possible. It is where we first encounter tired the fact that recent values are truly a little different than those previously determined, but that all contributed to the increase in the measuring precision, that is, the reduction of the mistakes in measurements. The absolute truth is that the measuring precision is increasing, but the 'constants' are changing a little and slowly.

I will go back to Einstein's STR. His postulate that the speed of light was a constant and that it was the most outstanding possible speed in nature, was wrong for several reasons: We have seen why and how c is not a constant, we have seen that there is an increase in the speed of light; therefore, its numeric value exceeds itself over time, our knowledge of nature is fragile now, let alone a century ago, so saying that the speed of light was the greatest in nature was not a reflection of human wisdom but a 'daily scientific need.' Einstein needed not to get a negative algebraic sign in the value under the square root. Then, the mathematical apparatus of the STR would be completely absurd and, as such, utterly unacceptable for physicists. I will remind you that Einstein's theories of relativity had several opponents from the beginning, and there are many today, too. That is why, in 1921, Einstein won the Nobel Prize not for his theories of relativity (due to which he became and remained famous) but for his explanation of the photo effect.

I believe that the entire process of introducing the constants into physics came and developed because of the 'daily scientific' needs. Here is a clarification of this statement.

When physics was, in the past times, performing experiments to determine the dependence of a value on several others, then they expressed that dependence mathematically, mainly by dividing some values' products by a product of some other values, and that, to be able to use that form- late in reality, introduces 'constants.' They were deciding the 'constants' from measurements, and that was it. The introduction of the 'constants' satisfied the 'daily scientific needs' to obtain applicable formulae without a complete understanding of what was examined and calculated by those formulae.

Logic tells us that the number of the introduced 'constants' is always by the level of our not understanding of the phenomena and the values we have been examining.

I agree that it might have been a necessary step in developing physics. 'Let's use something, even though we do not understand the real nature of that', was the expression of human pragmatism. However, technological 'advancement,' without understanding the essence, has led us to a situation where we have jeopardized our existence, Because we have jeopardized the functioning of planet Earth. It is high time we began to realize what is what, and by exerting heroic efforts, we tried to fix the damage we have caused so far.

And now more about the non-universality of the 'constats.

When I say that the 'constants' are not universal, I would like to say that the values that we have measured here on Earth will not be identical with quality the values we would measure at another spot in the cosmos. The quality of the space changes from one place to another.

The space is not homogenous. And when it is not homogenous, it is not isotropic. 'Isotropic' means that it is all the same for us in which direction we are moving or seeing.

Proof? Well, look at the sky. Is the allocation of the celestial bodies homogenous? No. Is it all the same in which direction we are looking? No.

I agree that we can say that a certain part of space, in a brief time interval, is homogenous and isotropic. We must be practical when it is demanded.

For The End Of The This Part

When I realized how huge the job I had started was and how much time it could consume, I decided to make partitions. So, what has been said so far will be the first part. I do not know how many parts there will be. The time will show.

I sincerely hope to get help in my work since the amount of work is enormous and long-lasting. I am but a pioneer, a starter, an initiator.

I also believe that in the era of information technology and the internet, spreading my ideas will be made quickly.

I expect a quick reaction to everything I have said whether positive or negative.

Innovative ideas always have a tough time getting through. I know that very well. But I do hope that the XXI century will justify and prove that it's quick and uncompromising, in my case, too.

I do not expect mercy or any kind of protection. I expect an argument and clash of opinions. I expect objectivity. I expect a true thirst for new knowledge and the truth.

In knowledge, there is no democracy. The truth does not depend on the voters' number that will support it. There is only one voice needed. But it is certain that the time of the acceptance depends on the number of voters and of their quality, too, because not all votes count equally.

If the time is ready for changes, and I personally think it is, then, everything will be easier.

I am personally sure that the XXI century will be marked by the development of new physics. When physics develops rapturously, it affects all other natural sciences. Modern technologies that might come out of general science development can and will change our future.

I hope we are conscious of the necessity of complete changes.

ABOUT AUTOR

I was born in 1963 in Nis City, Republic of Serbia, ex-Yugoslavia. I went to school in Nis, and here I live and work. After the "specialized training," I gained the title of laboratory technician for physics, and I enrolled myself in the studies of physics at the Department of Physics at the Faculty of Sciences in Nis, the University of Nis, where, after finishing it, I gained the title physicist. Even as a graduate, I discovered contemporary physics mistakes and drawbacks. My first autonomous scientific work, "Classical interpretation of the Michelson-Morley experiment" (70 A4 pages with several incredibly detailed drawings and adequate accompanying mathematics), did not meet understanding, neither with my colleague students nor my professors. Nobody replied, not in the country or abroad (I sent the English version to some addresses worldwide).

In 1990, when I was spiritually ready, my spiritual preacher Sri Chinmoy accepted me as his student; I was initiated into regular meditation and concentration on my spiritual heart.

However, my persistence paid off in 1999 with my second independent work, 'Mass temperature relativity, the secret of anti-gravity'. This groundbreaking piece, spanning 7 pages, was published the following year in the electronics magazine 'Journal of Theoretics'. The same year, I had the honor of presenting it at a scientific congress in Sankt Petersburg, Russia, marking a significant milestone in my career.

In 2003, I ventured into the world of inventions, a path leading to international recognition. The following year, I was awarded the bronze prize at the prestigious international fair 'Pal Expo' in Geneva, for my 'HSP motor.' In 2008, I authored the book 'THE INTRODUCTION INTO NEW PHYSIC, Part One ', a significant contribution to the field, albeit published only in Serbian. 2013 during my stay in Toronto, I began public presentations and YouTube promotions of the New Paradigm of the Universe, further solidifying my global reach and impact.

Since then, I have been researching and developing modern technologies to obtain pure energy from renewable sources.

Contact: goranmitic369@gmail.com

———

Ilustration
Saša Dimitrijević

Pictures Goran Mitić
Photos
from NASA and SOHO websites

9 798889 569055